THE MARINE FISHES OF NORTH-WESTERN AUSTRALIA
A GUIDE FOR ANGLERS AND DIVERS

95: Dives #1-6 and (snorkel) #7

6 thru 10 April 1995 around the NW cape based in Exmouth W.A. Dives #1, 2, 4 and (snorkel) 7 were in Northern Ningaloo Reef environments. #3, 5 were under the Navy Pier near the tip of the cape and #6 was to the east of the cape on Bundegi Reef.

A general guide to inshore fishes
of tropical Australia

THE MARINE FISHES

OF NORTH-WESTERN AUSTRALIA

A Field Guide for Anglers and Divers

Gerald R. Allen
Roger Swainston

Western Australian Museum
1988

PREFACE TO THE THIRD EDITION

Various corrections have been made to the previous edition involving the following species: 159, 316, 395, 527, 529 and 769. We also take this opportunity to acknowledge the Christensen Research Institute (Papua New Guinea) and its Director, Dr. Matthew Jebb, for allowing us to use the paintings of batfishes (Platax) in plate 38 that first appeared in the second edition.

First Published 1988
Revised Edition 1990
Second revised edition 1993
Reprinted 1994
© Western Australian Museum 1988
ISBN 0 7309 2113 1

Published by the Western Australian Museum
Francis Street, Perth, Western Australia 6000
Cover design by Lorinda Reincastle
Colour separations by Quadrascan Graphics, Perth, Western Australia
Printed by Scott Four Colour Print, Perth, Western Australia

CONTENTS

Figure 1. Map of Western Australia showing most of the localities mentioned in the text.

INTRODUCTION

Although the exact total is unknown it has been estimated there are nearly 22 000 types of fishes inhabiting our planet. Of this total approximately 13,500 are confined to marine environments. The seas surrounding the Australian continent are inhabited by the richest fish fauna on earth. Approximately 3400 species have thus far been recorded, an amazing 25 per cent of the world's total marine fishes. Our huge number of species is partially explained in terms of geography. Because the continent impinges on both tropical and temperate latitudes we essentially have two very distinct assemblages of fishes. In addition, there is a subtropical zone forming a buffer between the two which also contains distinctive fishes. It is a well known fact among zoogeographers (scientists who study the distribution of animals) that the richest marine biological province on earth in terms of diversity and sheer number of species is the Indo-Australian Archipelago. This tropical zone which encompasses northern Australia also includes the Malaysian Peninsula, Indonesian Archipelago, Philippines, and the islands of Melanesia. A number of theories have been put forward to account for the tremendous biological richness of the region. There is no doubt a combination of several factors responsible. Certainly some of the most important would be a relatively long history of favourable climatic conditions, a great diversity of ecological conditions, and a tumultous geological and hydrological past. In the latter category events such as sea-level changes, current changes, volcanism and shifting land masses have created isolating barriers that enhanced the process of speciation.

Perhaps the most significant factors responsible for Australia's great plethora of fishes is its vast tropical shoreline with an array of diverse habitats. First and foremost in this respect is the coral reef. The reef abounds with living space and what can be termed 'survival opportunities' for fishes and other organisms. Like multi-storey tenement buildings, spires of live tabular and branching corals rise above the surface of the reef. They offer shelter to a myriad of small, brightly coloured fishes that swarm above them. Ledges, caves and crevices form the inner sanctum of the reef city. Here a community of secretive fishes dwell, some which are seldom seen or only emerge for night-time feeding patrols. At first glance the sandy fringe that often surrounds individual reef complexes appears to be devoid of fish life. However, close inspection reveals another equally interesting community of specially adapted and often well camouflaged fishes, including many that burrow under the surface. The above reef habitats in addition to inshore coastal environments and offshore sectors of the continental shelf are each populated by distinct fish communities.

In spite of the wealth of marine fishes inhabiting our north-western seas, there has been a notable lack of published information. Aside from the recent works of Gloerfelt-Tarp and Kailola (1984) and Sainsbury *et al.* (1985) which deal primarily with trawl fishes of the North West Shelf, there is no comprehensive guide to coastal fishes. This book is therefore designed to fill this need. It is the end result of 14 years of continuous diving, collecting, and study by the authors and incorporates the research of numerous colleagues who have studied the classification of north-western fishes.

The book contains more than 1200 hand-painted illustrations of 1062 species. The paintings were completed over a 15-month period and are primarily based on photographs or colour transparencies of either live fishes taken underwater or freshly caught specimens. In many cases preserved specimens at the Western Australian Museum have been consulted to ensure accuracy of detail and proportions. The end result is a colourful and highly comprehensive guide to the sea fishes of the North-West. It represents an ideal companion volume to the guidebooks dealing with southern Australian fishes by Hutchins and Thompson (1983) and Hutchins and Swainston (1986).

AREA OF COVERAGE

This book is intended to give comprehensive cover of the reef and shore fishes living along the coast from Shark Bay, Western Australia northwards to about Darwin. It is also reasonably comprehensive for the area east of Darwin to Cape York. Virtually all of the common inshore species are included. Full species coverage is given to most families of reef fishes except the gobies (Gobiidae) and threefins (Tripterygiidae). Both of these families contain small, cryptic, seldom seen species, and the classification of both groups has not been satisfactorily studied; many of the species are difficult to identify, even by trained specialists.

Full coverage is also extended to families containing species that are of interest to anglers. Foremost in this respect are the trevallies and their relatives (Carangidae), tunas and mackerels (Scombridae), and billfishes (Xiphiidae and Istiophoridae). Less than full coverage is given to families containing cryptic, hence seldom encountered species, or species living in deeper sections of the continental shelf. An estimated 70 per cent of the bottom dwelling fishes of the deeper continental shelf (to about 100-200 m) are represented. Excluded from coverage are the true deep sea fishes which for the most part, live well offshore, below 200 m depth (although some make daily migrations to the surface). The fishes from this realm in Western Australia have scarcely been collected or studied.

In summary, this book is intended to serve as a field identification guide to the marine and estuary fishes of north-western Australia. Nearly all species are included that will be encountered by anglers, trawl fishermen, SCUBA divers and snorkelers. Less than full coverage is given for the deeper offshore fauna that is primarily of interest to research scientists. Coral reefs situated far offshore on the outer edge of the North West Shelf including the Rowley Shoals, Seringapatam Reef, Scott Reef, and Ashmore-Cartier Reefs are outside of the area of coverage, although many of the fishes occurring there are included in the book.

THE NORTH-WESTERN FISH FAUNA

The fish fauna of Australia is composed of two separate and very distinct groups which are correlated with sea temperatures. The southern fauna containing about 600 inshore species inhabits the coast between Shark Bay, Western Australia and Moreton Bay, Queensland. Although many of the genera and most families also occur in northern Australian seas, nearly all of the species are unique. In fact most are restricted to southern Australia, although a number of them also occur in New Zealand.

Slightly over 2000 inshore fish species have thus far been recorded from Australia's northern half. In contrast to the fishes in cool southern seas, a majority of the tropical species are widely distributed in the vast Indo-Pacific region. Indeed, many of these fishes exhibit distribution patterns extending from the shores of eastern Africa to the islands of Micronesia and Polynesia. The dispersal of these wide-ranging fishes is facilitated by their possession of pelagic egg and larval stages which float on or near the sea's surface at the mercy of waves, winds, and currents.

In between the zones containing tropical and temperate fishes there are transition areas containing a mixture of northern and southern fishes, as well as a number of unique species. On the west coast this transition zone is situated between about Lancelin (31°01'S) and Point Cloates (22°43'S). A few examples of fishes that have their main breeding populations within this area are the West Austalian Jewfish *(Glaucosoma hebraicum)*, Western Buffalo Bream *(Kyphosus cornelli*, see No. 596), and Baldchin Groper *(Choerodon rubescens*, see No. 739).

The main area of coverage in this book lies immediately to the north of the western transition zone. It stretches for about 5000 km to the coast of Arnhem Land. Although much of the coastal scenery is arid desert, the offshore waters have much to offer for anglers, naturalists and scientists. It includes Australia's longest inshore barrier reef, stretching some 250 km between Cape Farquhar and North West Cape. Farther northwards there is substantial reef development around offshore islands such as the Murion Islands, Barrow Island, the Monte Bello Islands, and the Dampier Archipelago. Between the last mentioned locality and Derby the coastal waters consist mainly of extensive flat sandy areas with few reefs. Mudflat and mangrove swamps are also common at a number of localities between Exmouth Gulf and Derby. The Kimberleys and adjacent Northern Territory represent the northernmost outposts of the north-west region and offer some of the finest coastal scenery on the continent. Spectacular cliffs of red sandstone form ramparts along the shore and extensive mangrove estuaries occur at river mouths. Coral reefs are also present around some of the offshore islands, but diving and fishing are difficult due to reduced visibility and excessive tidal currents.

The north-western region is inhabited by approximately 1400 species of fishes that occupy a wide diversity of habitats. Those which are included in this book range seaward to the deeper parts of the continental shelf. Another entirely different assemblage of fishes lives beyond this area on and above the continental slope, an area of which we know very little. Research on this portion of our fish fauna has until now been seriously hampered by the lack of adequate research vessels. Hopefully some day we will know as much about these fascinating creatures as we do about the more familiar inshore fishes.

ACKNOWLEDGEMENTS

We are most grateful to the Director and Board of Trustees of the Western Australian Museum for their support of this project. Details of planning, typesetting, and technical layout were facilitated by Museum Publications Department staff members Ann Ousey, Ione de Alwis and Greg Jackson.

We thank Barry Hutchins for his useful ideas and comments about the manuscript. The general format of this book is based largely on that of Barry Hutchins' and Roger Swainston's *Sea Fishes of Southern Australia* and is intended to serve as a companion volume.

We are also indebted to Nick Haigh of the Museum's Department of Ichthyology for attending to a large part of the department's demands and services during the period when the book was being prepared.

Field work in north-western Australia was greatly assisted by the following people: Tony and Avril Ayling, John Braun, Norrie Cross, Eve and Bill Curry, Ian Parker, Neil Sarti, and Barry Wilson. I am particularly grateful for the assistance and companionship of the Museum's marine biological group including Paddy Berry, Clay Bryce, Louisette Marsh, Gary Morgan, Shirley Slack-Smith, and Fred Wells.

Numerous scientists in Australia and overseas contributed to our knowledge of our state's fishes either through their publications or direct assistance with problematical identifications. Several who were particularly helpful in this respect included Tony Gill, Doug Hoese and John Paxton (Australian Museum), Rudie Kuiter and Martin Gomon (Museum of Victoria), Barry Russell and Helen Larson (Northern Territory Museum), Rolly McKay (Queensland Museum), Ronald Fricke (Natural History Museum, Braunschweig, W. Germany), William Eschmeyer and John McCosker (California Academy of Sciences), Ed Murdy, Jeff Williams, and Victor Springer (Smithsonian Institution), Theodore Pietsch (University of Washington), Stuart Poss (Gulf Coast Research Lab, U.S.A.), Jack Randall (Bishop Museum, Honolulu), Bill Smith-Vaniz (Philadelphia Academy of Sciences), Richard Winterbottom (Royal Ontario Museum), and Peter Whitehead (British Museum, Natural History).

Finally we wish to acknowledge Keith Sainsbury, Patricia Kailola, and Guy Leyland for their fine publication *Continental Shelf Fishes of Northern and North-western Australia* containing colour photographs that served as models for many of Roger Swainston's paintings. Similar assistance was provided by *Trawled Fishes of Southern Indonesia and Northwestern Australia* by Thomas Gloerfelt-Tarp and Patricia J. Kailola.

CLASSIFICATION OF FISHES

Although the fundamentals of biological nomenclature and classification are common knowledge to many, it is my experience that the average non-biologist frequently has little idea of the basis of scientific names or how fishes are classified. It therefore seems worthwhile to include a brief section on the rudiments of this subject.

Every described organism, be it a single celled amoeba, crab, bird, fish or mammal has a scientific or Latin name. It is composed of two parts and is generally italicised. The first part is the genus or generic name and the second is the species or specific name. For example the Five-lined Seaperch is *Lutjanus quinquelineatus*. The generic name *Lutjanus* pertains to a group of closely related species which share a number of common features related to general shape, scalation, type of teeth, fin-ray counts, etc. The specific name *quinquelineatus* applies only to a single entity that is distinguished from its relatives by a unique set of characteristics, often including colour pattern. Related genera (plural of genus) are grouped together in a family, whose spelling always ends in - idae. An illustrated list of families is presented on pages 9-18. Worldwide there are 445 families; 300 are represented in Australia. A group of similar families is placed in one of the 35 orders of fishes whose spelling always ends in - iformes. The highest rungs on the 'ladder' of classification pertain to class and phylum. The class Myxini contains the jawless hagfishes and lampreys (no species included in this book); Chondrichthyes contains sharks and rays; and the third class Osteichthyes contains the majority of fishes. All fishes, as do other higher animals including amphibians, reptiles, birds and mammals, belong to the phylum chordata. Therefore, in summary the classification of the Five-lined Seaperch can be represented as follows.

Phylum - Chordata (all animals with notochord)
Class — Osteichthyes (all bony fishes)
Order — Perciformes (most reef fishes)
Family — Lutjanidae (seaperches and relatives)
Genus — *Lutjanus* (closely related seaperches)
Species — *quinquelineatus* (5-lined seaperch)

Characters that are most often used to separate species, and often genera, include external features such as the number of fin rays, size and number of scales, ratio of various body proportions, and colour pattern. For higher classification at levels above genus internal structure, particularly those pertaining to skeletal elements, are often indicative of relationships.

Many species previously unknown to science have been found in Western Australian seas over the past few decades. When a new fish is discovered it is given a scientific name by the researcher who formally publishes a detailed description in a recognised scientific journal. Scientific names are frequently descriptive. For example *quinquelineatus* is Latin for five lines and is therefore appropriate for the Five-lined Seaperch (see No. 499 in species section). New fishes are sometimes named after the locality from where they are collected, for example *japonicus* (Japan) or *novaeguineae* (New Guinea). A third category of specific names are based on the names of people, often the person who first discovers the fish (respectable reseachers never name fishes after themselves). Fishes named after a male end in - i, those after females in - ae.

PRESERVING FISHES

It is sometimes desirable to preserve specimens, particularly if a positive identification by museum authorities is required. Also small, unusual or rare fishes can be kept as curios or as teaching aids for children. The recommended method of preservation in any case is exactly the same one that is employed by fish biologists in museums. The basic ingredient is full strength formalin which can be obtained from a pharmacy. The preserving solution is made by diluting one part of formalin with nine parts of water. The fish should be fully immersed in the solution. If larger than about 15-20 cm a slit along the side of the belly will facilitate preservation of the internal organs. For long term storage it is desirable to transfer the specimen to a 70 per cent ethyl alcohol solution (70 per cent ethanol, 30 per cent water) after the fish is fully fixed in formalin (i.e. after several weeks). However, the fish may be held in the initial formalin solution for several years without deleterious effects.

Unfortunately colours fade rapidly in preservative. Therefore photography (see below) is a valuable method of accurately recording the colour pattern.

SENDING SPECIMENS TO THE MUSEUM

Although most of the specimens in the reference collections of the various state museums around Australia are collected on special expeditions by museum staff, occasionally valuable fishes are donated by the public. Also it may be desirable for people living far from their local museum to send specimens in for identification, particularly if the fish in question is suspected to represent a new record for the area or perhaps is very rare. Also members of the public may have the opportunity to collect fishes in remote areas that are not easily reached by museum scientists. For example, several years ago, a medical officer aboard an experimental offshore drilling platform obtained a valuable collection of deep reef fishes on the North West Shelf that were accidently captured when the drill was brought up from 120 m depth. In this case none of the fishes were recognised by the crew which included several anglers, so the specimens were wisely preserved and sent to the Western Australian Museum. Several species from this collection proved to be previously unknown to science.

Specimens can easily be sent to museums via parcel post if first properly preserved (see above section). They should be removed from the preserving solution, rinsed, and wrapped in moist cloth (cheesecloth is ideal) or newspaper, then sealed in several layers of plastic bags. The bags can then be posted in a well padded cardboard box. Fishes from our north-western seas can be sent to either of the following institutions depending on their state of origin: (1) Department of Ichthyology, Western Australian Museum, Francis Street, Perth, W.A. 6000, or (2) Department of Ichthyology, Northern Territory Museum, P.O. Box 4646, Darwin, N.T. 5794.

FISH PHOTOGRAPHY

Nearly everyone carries a camera on fishing and diving expeditions these days. Good photographs can be valuable in determining the identification of a questionable fish, particularly if the catch has already been eaten. Anglers frequently bring photos to us at the Museum for identification. Their usefulness is sometimes diminished because little care was taken in preparing the fish. The following steps will ensure the photos are of good diagnostic quality: (1) The specimen should be photographed when fresh as live colours fade rapidly after death. (2) An attempt should be made to spread out the fins. With small fishes you can hold the fins erect with sewing pins on a piece of flat styrofoam or cardboard. If full strength formalin (available from pharmacies) is then applied to the fins with a small paintbrush or eye dropper and allowed to set for a few minutes they will remain erect when the pins are removed. (3) Wet fish should be blotted dry with a cloth or paper to prevent harsh glare when photographed. (4) The specimen should be placed on a suitable contrasting background and photographed as close as the lens will allow for sharp focus, attempting to fill the frame. (5) It is helpful if a ruler or some other object of known length can be placed besides the fish when it is photographed in order to determine its length later on.

Underwater photography is a fascinating hobby and will add a new dimension to your diving activities. Fish photography, if done on a regular basis, is an excellent method of learning the fishes of an area. Most beginners start out with a Nikonos or one of the relatively inexpensive automatics in a perspex housing. However, to obtain high quality fish portraits it is advisable to use an SLR camera housed in a special case made of perspex or aluminium alloy. In addition, strobe lighting is a must. The cost of the basic outfit ranges from about $3000-$5000, so only the serious photographer will consider this alternative. Even for accomplished divers it requires much practice and patience before good results are obtained. The combination of a moving subject on variable backgrounds presents a great challenge.

DANGEROUS FISHES

Although our seas are generally safe for normal swimming and wading activities there are a number of fishes potentially capable of causing injury. Several specialised books dealing with this subject (see *Dangerous Fishes of Western Australia* by J.B. Hutchins, published by the Western Australian Museum, 1980) are available so its treatment here will be brief. Additional detail is provided in the 'boxes' which appear opposite the fish plates.

Potentially harmful fishes can be divided into several broad categories including species that can bite, species which can sting, and species which can cause poisoning if consumed.

Biters — first and foremost in this category are the whaler sharks and their relatives (17-42). In addition there are a number of smaller reef fishes which, although they pose no threat to swimmers, can inflict painful bites if handled carelessly by anglers. For example, barracudas (719-722), razorfishes (785-786) and triggerfishes (Plate 66) are notorious in this respect. As a rule of thumb any fish with large, obvious teeth should be handled with care.

Stingers — virtually any fish which possesses rigid fin spines is capable of inflicting wounds if handled carelessly. Most are non-venomous and can be treated in the same manner as any puncture wound. Surgeonfishes (Plate 61) are equipped with scalpel-like spines that are either fixed in an erect position or fold into a groove along the base of the tail. Spearfishermen in particular, need to exercise special care when removing these fish from spears as large specimens can sever a finger. The most dangerous category of stingers includes fishes which have venomous spines. The best known of these are stingrays (Plates 4-5), catfishes (Plate 9), scorpionfishes (Plates 15-16), and spinefeet (Plate 60). For all of these fishes the recommended first aid procedure is to immerse the injured area in hot water (as hot as bearable), repeating until the pain subsides. Apparently the protein base of the toxin is denatured by heat and relief is sometimes immediate. In cases where the victim is stung by several spines, or if the wound is deep, medical assistance should be obtained. Lionfishes (232-233), firefishes (234-238), stonefish (239), and stingers (246-247) have very potent venom in all fin spines. Several deaths have occurred outside of Australia as a result of people treading on the Estuarine stonefish (239) and failing to receive immediate first aid.

Poisonous fishes — there are two main types of fishes in this category. The first includes species that have naturally occurring poisons either in their external mucus or in some internal organs, frequently the viscera or gonads. The best known examples are pufferfishes, porcupinefishes, and boxfishes (Plates 68 and 69). Although these fishes are eaten by the Japanese when specially prepared by licensed chefs, they are considered extremely dangerous and specimens from local waters should never be eaten. The symbol Ⓟ is used in the species accounts to indicate fishes that are naturally poisonous. The second group of poisonous fishes includes species that acquire toxic properties during their life cycle by accumulating a dinoflagellate that lives on dead coral or among algae and is first consumed by herbivorous fishes which are eventually eaten by larger predatory fishes. The toxin known as ciguatera is accumulative and the largest fishes such as the Red bass (492) and barracuda (720) are potentially the most dangerous. The symptoms from eating ciguatoxic fish appear from one to 10 hours later and range from mild dizziness, diarrhoea, and a numb sensation of the lips, hands, and fingers, to extreme nausea, coma, and total respiratory failure. The degree of poisoning depends on the amount of fish that is consumed and the concentration of toxin it contains. Fortunately very few cases of ciguatera have been reported from Australian seas. As a matter of safe practice it would be wise to avoid eating large barracuda, Red bass, or large groupers (Plates 18-20), all of which have been implicated in ciguatera poisonings in other regions.

FISH VS. FISHES

Confusion is frequently expressed over the use of the words fish and fishes. The term 'fish' in particular is often used inappropriately. It is grammatically correct to use fish when referring to a single individual or more than one individual if only a single species is involved. For example, one might say 'there were 100 fish in that school of Spanish mackerel'. The term fishes is a plural form that is used when referring to two or more different species. For example, 'we saw hundreds of fishes from the glass-bottom boat'.

HOW TO USE THIS BOOK

This book is designed as a pictorial guide that relies on visual comparison between the painted illustrations and actual specimens, photographs, or underwater observations. Distinguishing features are highlighted in the text accompanying each plate, in most cases referring to colour pattern or the shape of the body or fins. These features are useful for differentiating the species in question from its close relatives, or species it is likely to be confused with. A guide to families based on outline drawings precedes the species section. When attempting to place a fish in the proper family particular attention should be given to the head and body shape, number of dorsal fins, placement of fins and their positions relative to one another, and presence or absence of spiny elements in dorsal and and anal fins in particular. The use of technical scientific words is deliberately avoided, but a few terms relating to the external features of fishes are useful for identification and are illustrated in Figure 2.

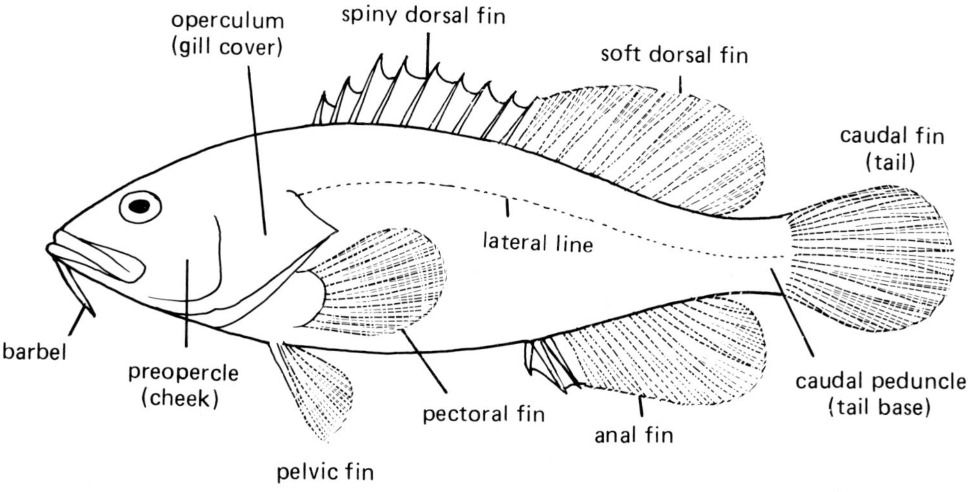

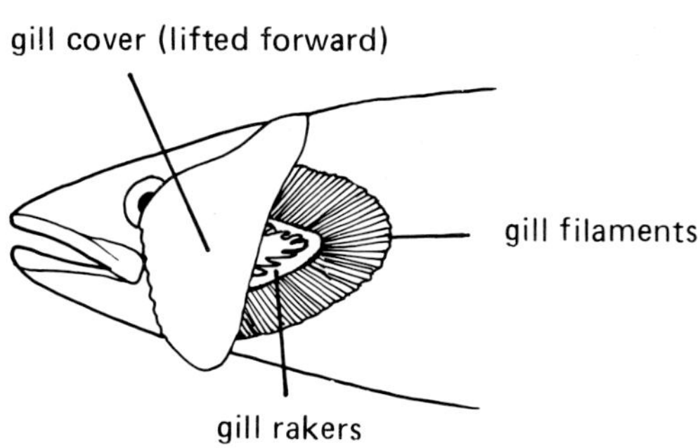

Figure 2. Diagram of a 'typical' fish showing external features.

Each species account appearing on the page opposite the corresponding plate includes the accepted common name and scientific name followed by the name of the person who first described it. If the person's name appears in parentheses it indicates that the species was originally placed in a genus different from its present one.

The text for each species contains general information on habitat, distinguishing features, and distribution, both within Western Australia and outside. Distributional limits are constantly changing and those which are given are merely intended to serve as a general guide.

The maximum known total length, measured from snout tip to the end of the tail, is given at the end of each species account. For many species this information is followed by a weight in kilograms which was the accepted Australian angling record at the end of 1987. These records are, of course, subject to change and anglers are urged to consult their local club officials for up-to-date information.

I have purposely omitted information on relative abundance (i.e. common, rare, etc.) as this parameter is subject to considerable local variation depending on availability of suitable habitat and in the case of migratory species, the time of year.

In addition to the individual species accounts, 'boxes' of text are included for most plates which contain general information for families, pertaining to such topics as number of species worldwide, ecology including food habits, and any noteworthy behavioural or morphological characteristics.

COLOUR PATTERNS

One shortcoming of all field guides is that is is virtually impossible to illustrate all of the variations in colour that commonly occur within a single species. With a few exceptions the colours shown here are the 'normal' or average ones displayed by live fish in their natural habitat. Anglers especially will be well aware that many fishes can drastically alter their colouration after being caught. Variation in colour pattern within individual species may also be related to age, sex, environmental conditions, or geography. Angelfishes (Plate 41), damselfishes (Plates 42-44), wrasses (Plates 47-51) and parrotfishes (Plates 52-53) are particularly notorious for dramatic changes in livery between the juvenile and adult stages. Wrasses and parrotfishes are also well known for their often different male and female patterns. Mainly due to budgetary restrictions it was not possible to illustrate all the variations related to sex and age, but they are included for a number of the more common species.

EDIBILITY RATINGS

The same system of edibility ratings used in Hutchins' and Swainston's *Sea Fishes of Southern Australia* is used in this book. Star symbols appear at the bottom right of each species account unless the fish is too small for human consumption or if there is no available information. Symbols are as follows. ★ = poor eating; ★★ = fair eating; ★★★ = good eating; ★★★★ = excellent eating; Ⓟ = poisonous. The star symbols are intended as approximate guides only. Wide variation in the edibility of a given species may be caused by a number of factors of which, degree of freshness and method of preparation are particularly important.

GUIDE TO FAMILIES

The following pages contain outline drawings of typical members of the families contained in the book. Scientific family names are indicated below each drawing and the Plate number is given in parentheses.

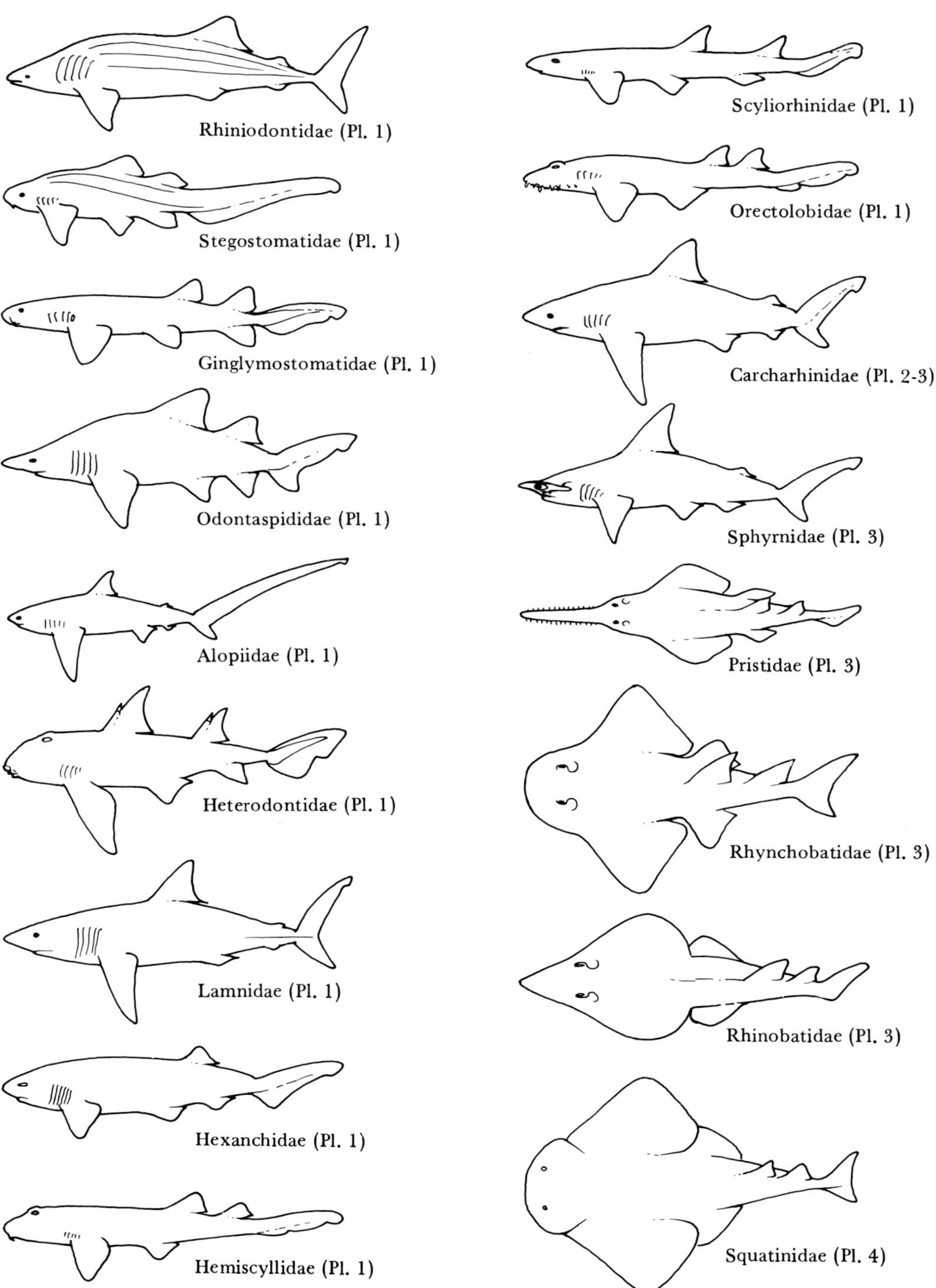

Rhiniodontidae (Pl. 1)

Stegostomatidae (Pl. 1)

Ginglymostomatidae (Pl. 1)

Odontaspididae (Pl. 1)

Alopiidae (Pl. 1)

Heterodontidae (Pl. 1)

Lamnidae (Pl. 1)

Hexanchidae (Pl. 1)

Hemiscyllidae (Pl. 1)

Scyliorhinidae (Pl. 1)

Orectolobidae (Pl. 1)

Carcharhinidae (Pl. 2-3)

Sphyrnidae (Pl. 3)

Pristidae (Pl. 3)

Rhynchobatidae (Pl. 3)

Rhinobatidae (Pl. 3)

Squatinidae (Pl. 4)

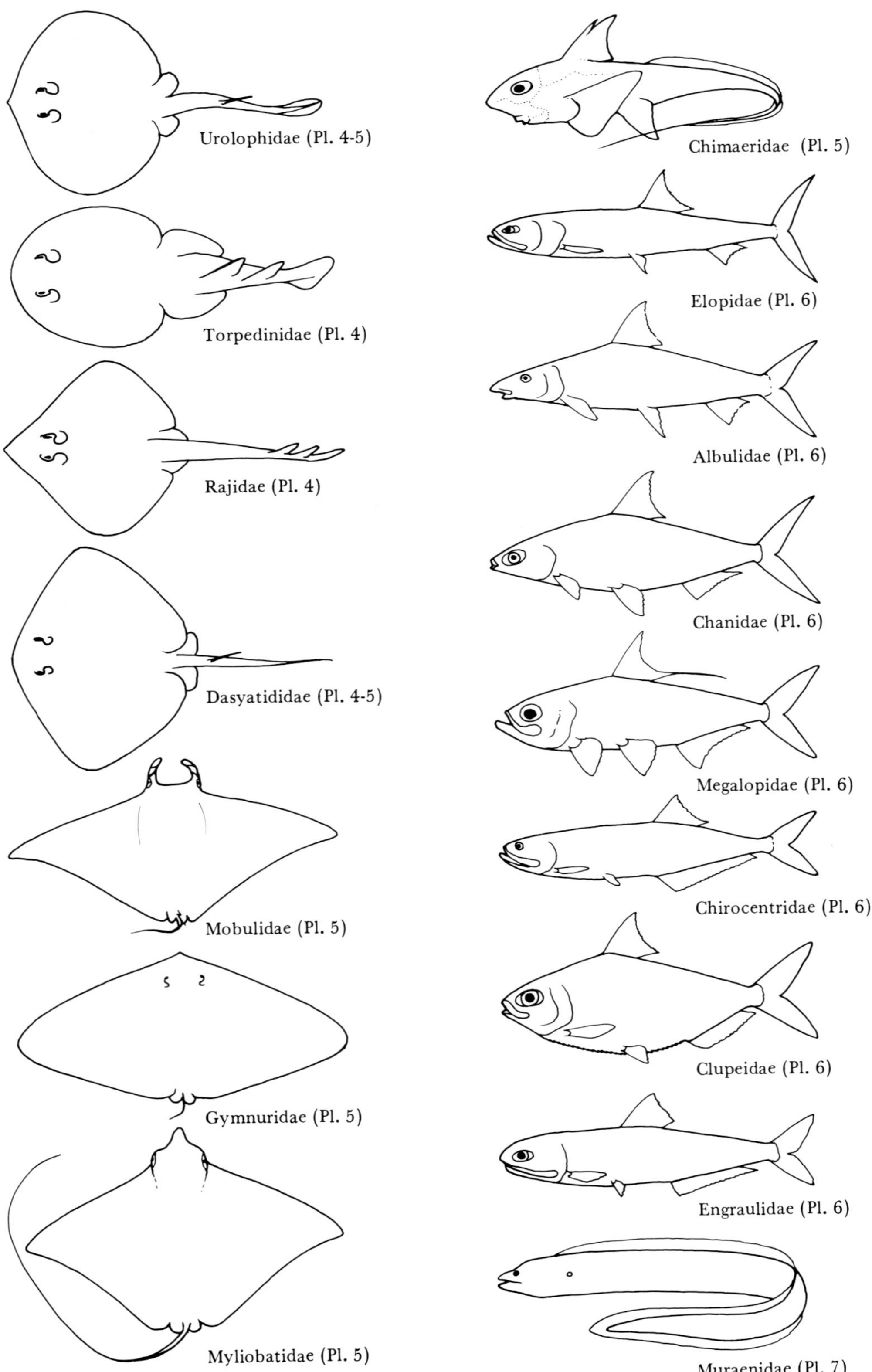

Urolophidae (Pl. 4-5)
Torpedinidae (Pl. 4)
Rajidae (Pl. 4)
Dasyatididae (Pl. 4-5)
Mobulidae (Pl. 5)
Gymnuridae (Pl. 5)
Myliobatidae (Pl. 5)
Chimaeridae (Pl. 5)
Elopidae (Pl. 6)
Albulidae (Pl. 6)
Chanidae (Pl. 6)
Megalopidae (Pl. 6)
Chirocentridae (Pl. 6)
Clupeidae (Pl. 6)
Engraulidae (Pl. 6)
Muraenidae (Pl. 7)

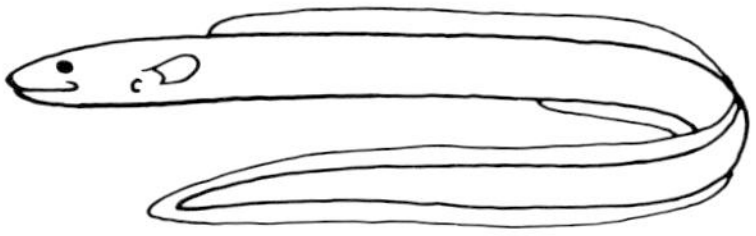

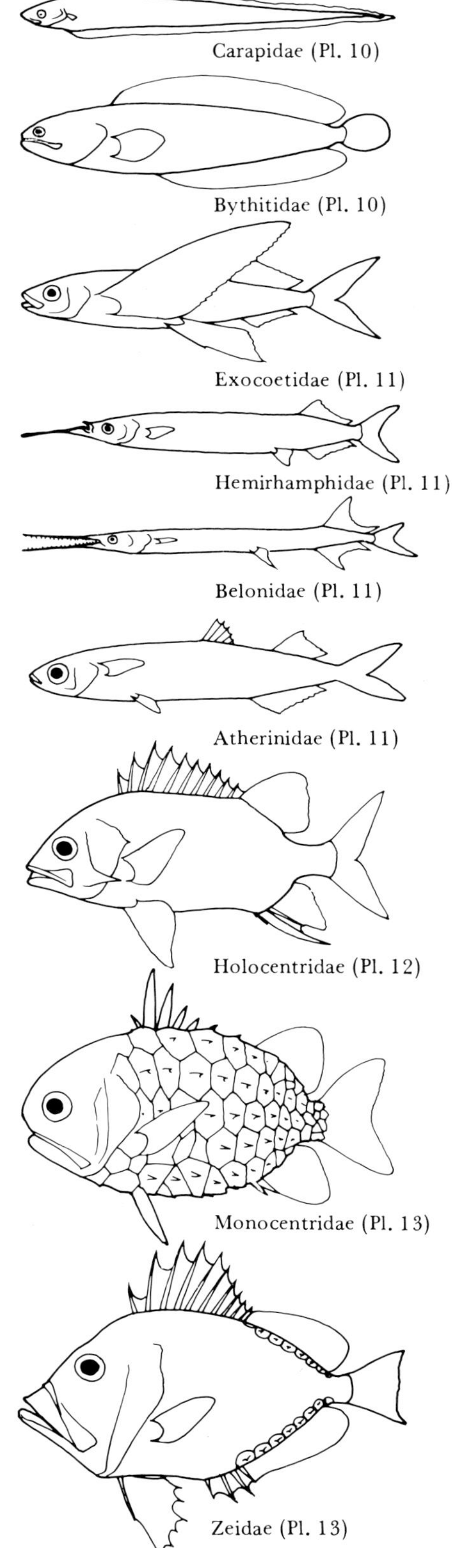

Congridae (Pl. 8)
Ophichthidae (Pl. 8) 
Ariidae (Pl. 9) 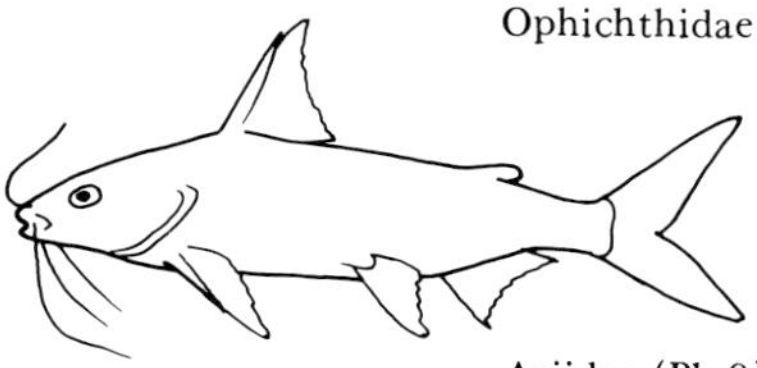
Plotosidae (Pl. 9) 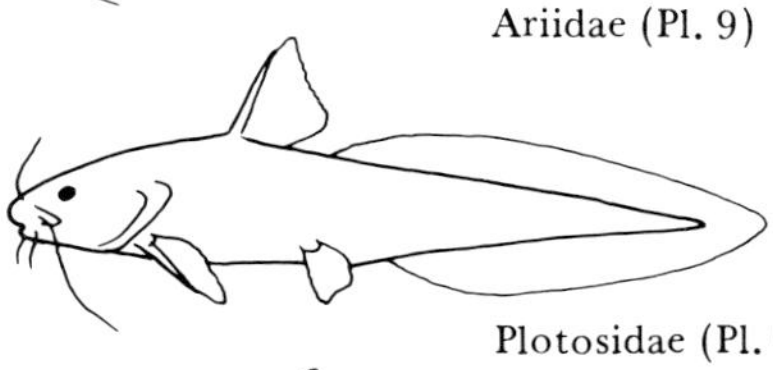
Synodontidae (Pl. 9) 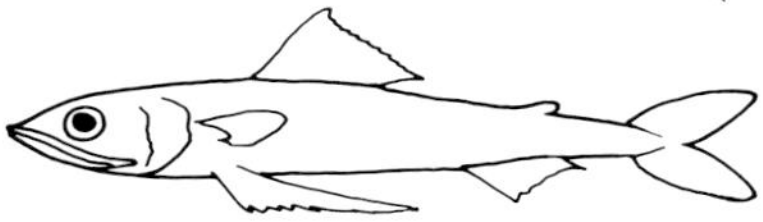
Harpodontidae (Pl. 9) 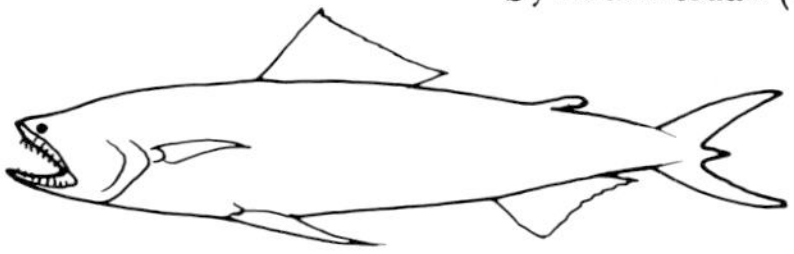
Batrachoididae (Pl. 10) 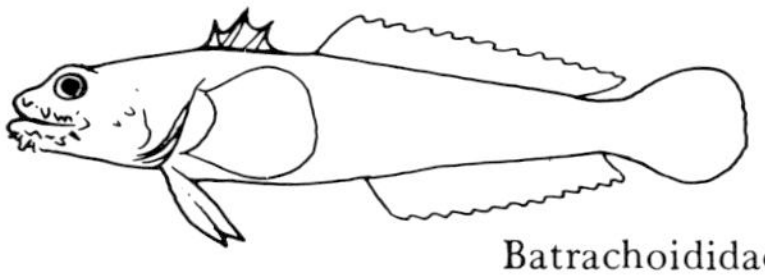
Gobiescosidae (Pl. 10) 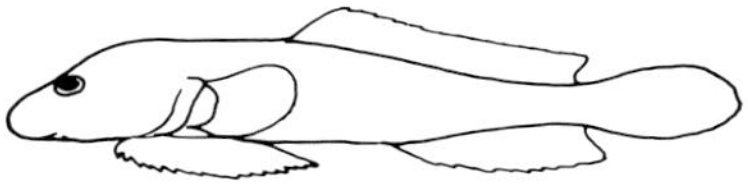
Antennariidae (Pl. 10) 
Ophidiidae (Pl. 10) 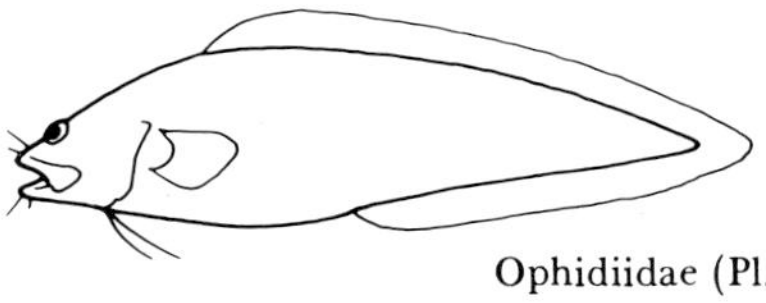
Carapidae (Pl. 10)
Bythitidae (Pl. 10)
Exocoetidae (Pl. 11)
Hemirhamphidae (Pl. 11)
Belonidae (Pl. 11)
Atherinidae (Pl. 11)
Holocentridae (Pl. 12)
Monocentridae (Pl. 13)
Zeidae (Pl. 13)

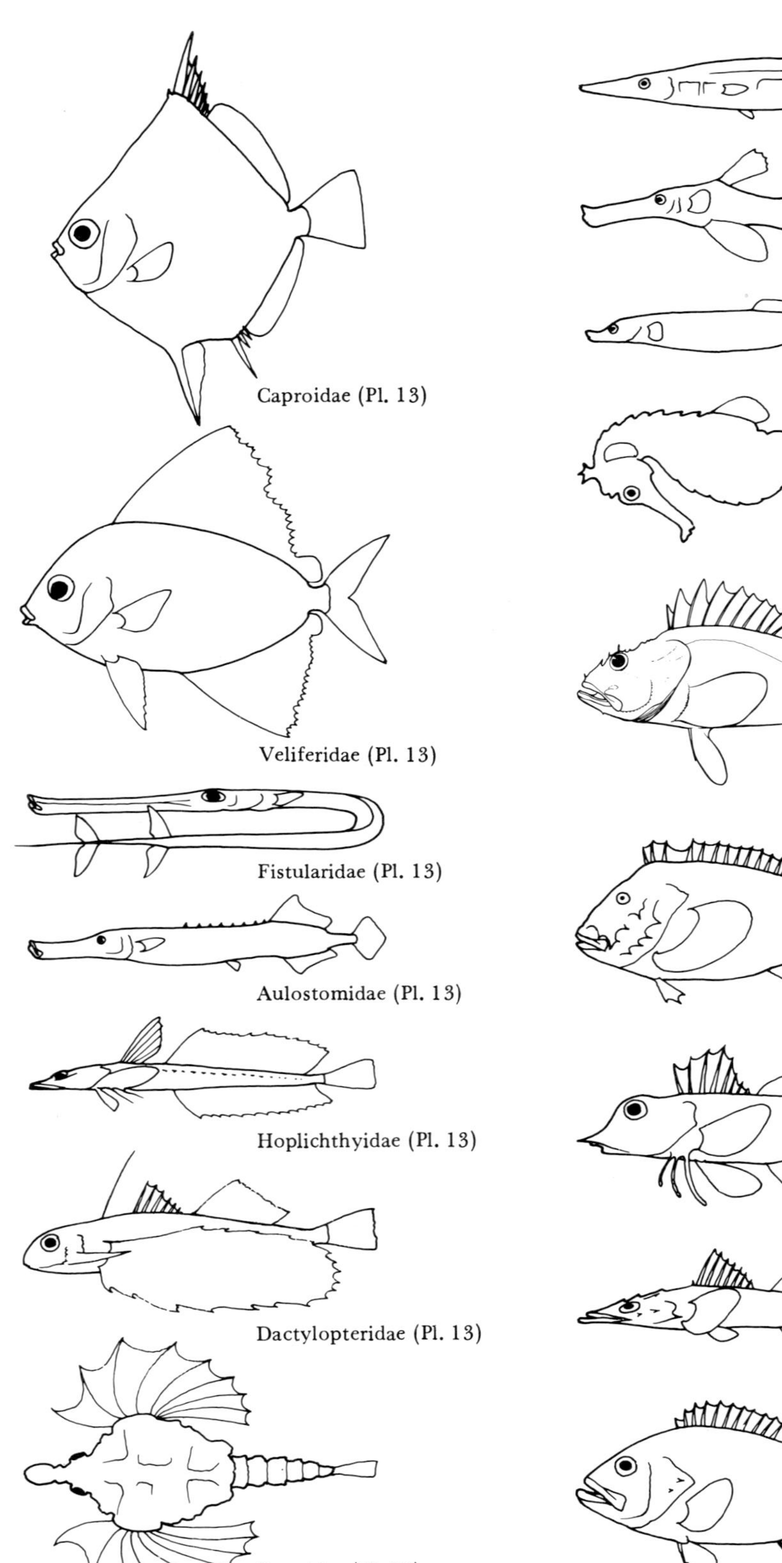

Caproidae (Pl. 13)
Veliferidae (Pl. 13)
Fistulariidae (Pl. 13)
Aulostomidae (Pl. 13)
Hoplichthyidae (Pl. 13)
Dactylopteridae (Pl. 13)
Pegasidae (Pl. 13)
Centriscidae (Pl. 14)
Solenostomidae (Pl. 14)
Syngnathidae (Pl. 14)
Syngnathidae (Pl. 14)
Scorpaenidae (Pl. 15-16)
Aploactinidae (Pl. 17)
Triglidae (Pl. 17)
Platycephalidae (Pl. 17)
Serranidae (Pl. 18-21)

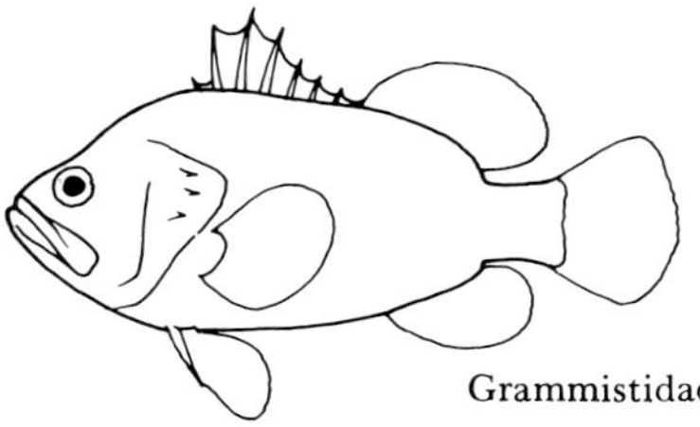

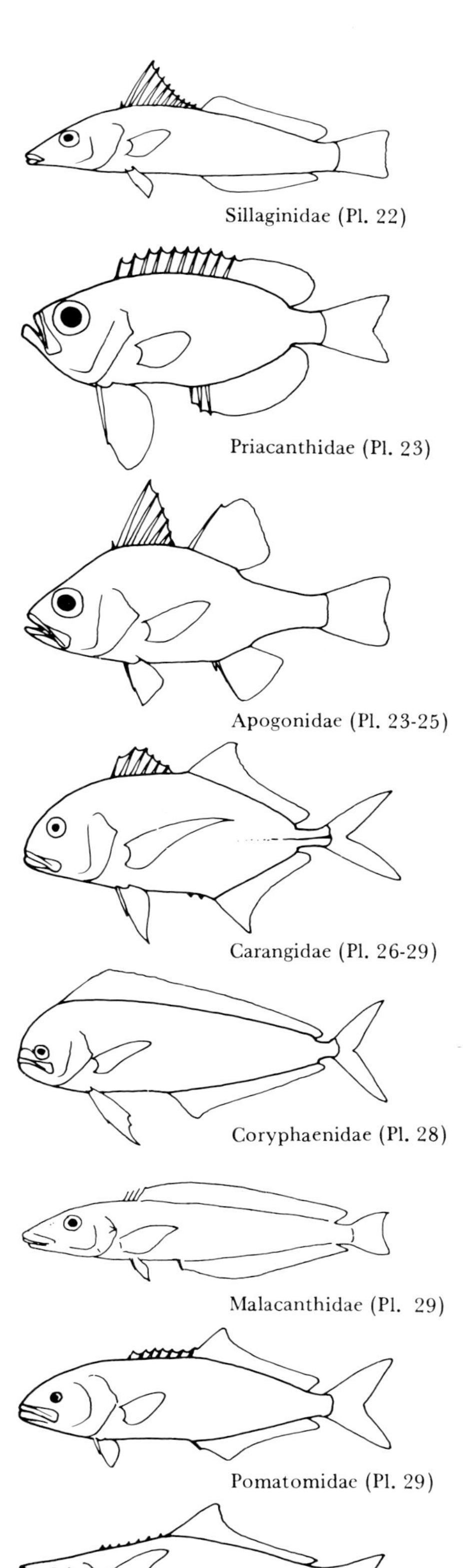

Grammistidae (Pl. 21)

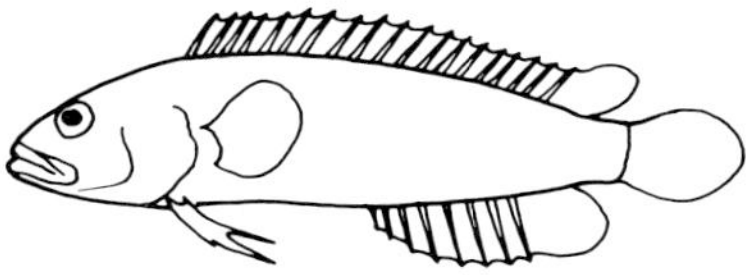

Acanthoclinidae (Pl. 21)

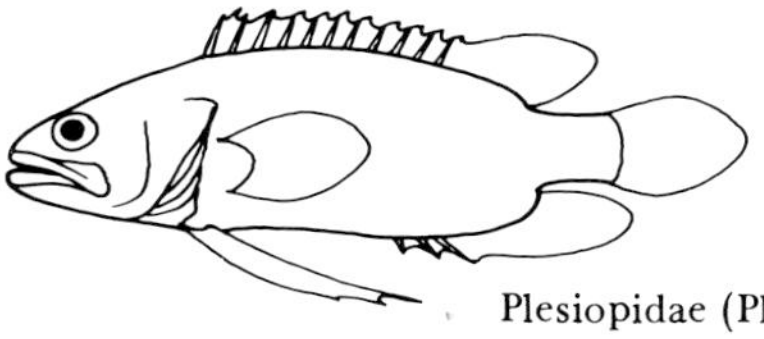

Plesiopidae (Pl. 21)

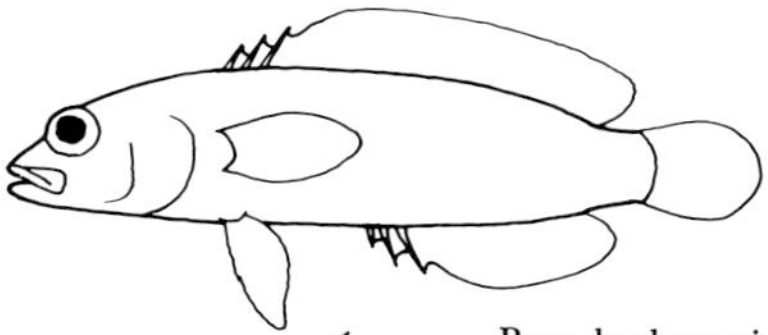

Pseudochromidae (Pl. 21)

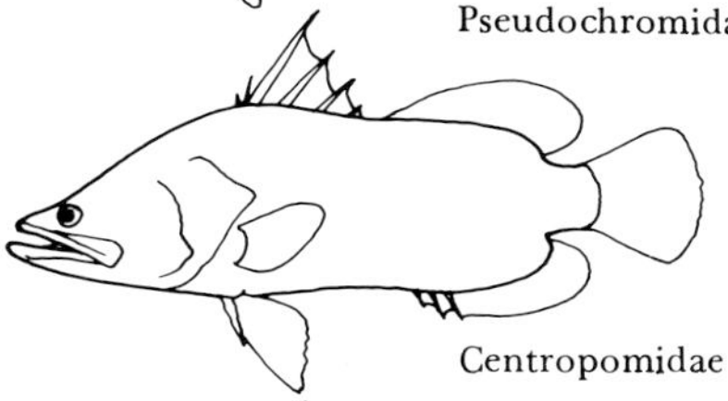

Centropomidae (Pl. 22)

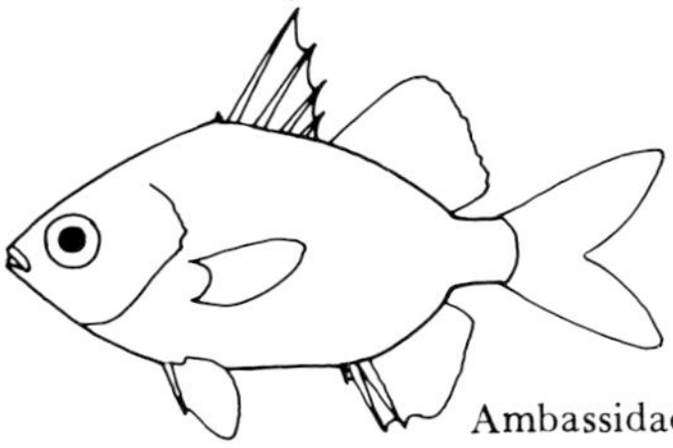

Ambassidae (Pl. 22)

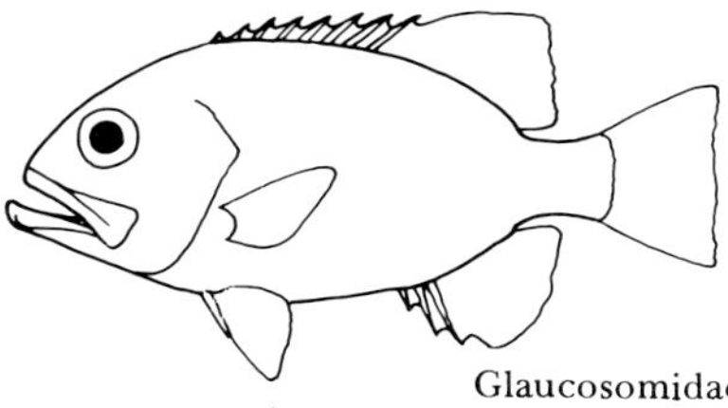

Glaucosomidae (Pl. 22)

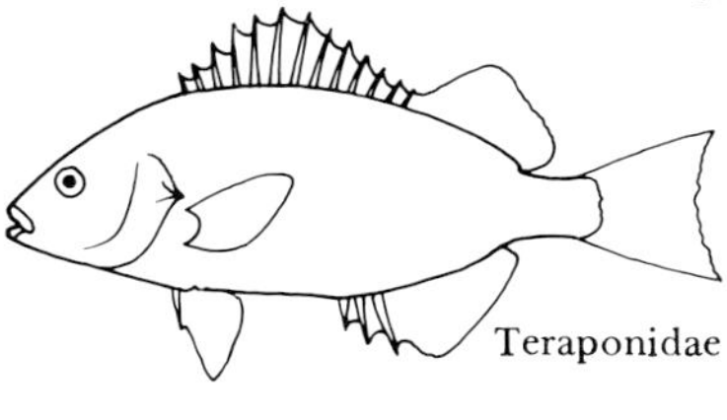

Teraponidae (Pl. 22)

Sillaginidae (Pl. 22)

Priacanthidae (Pl. 23)

Apogonidae (Pl. 23-25)

Carangidae (Pl. 26-29)

Coryphaenidae (Pl. 28)

Malacanthidae (Pl. 29)

Pomatomidae (Pl. 29)

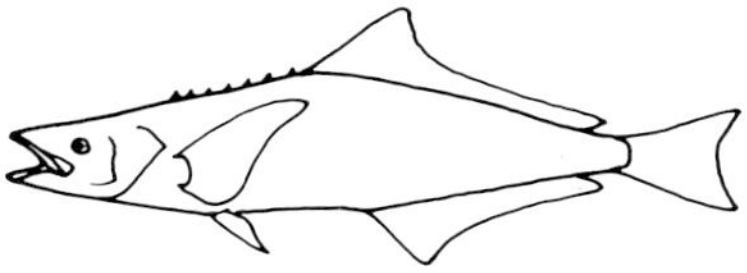

Rachycentridae (Pl. 29)

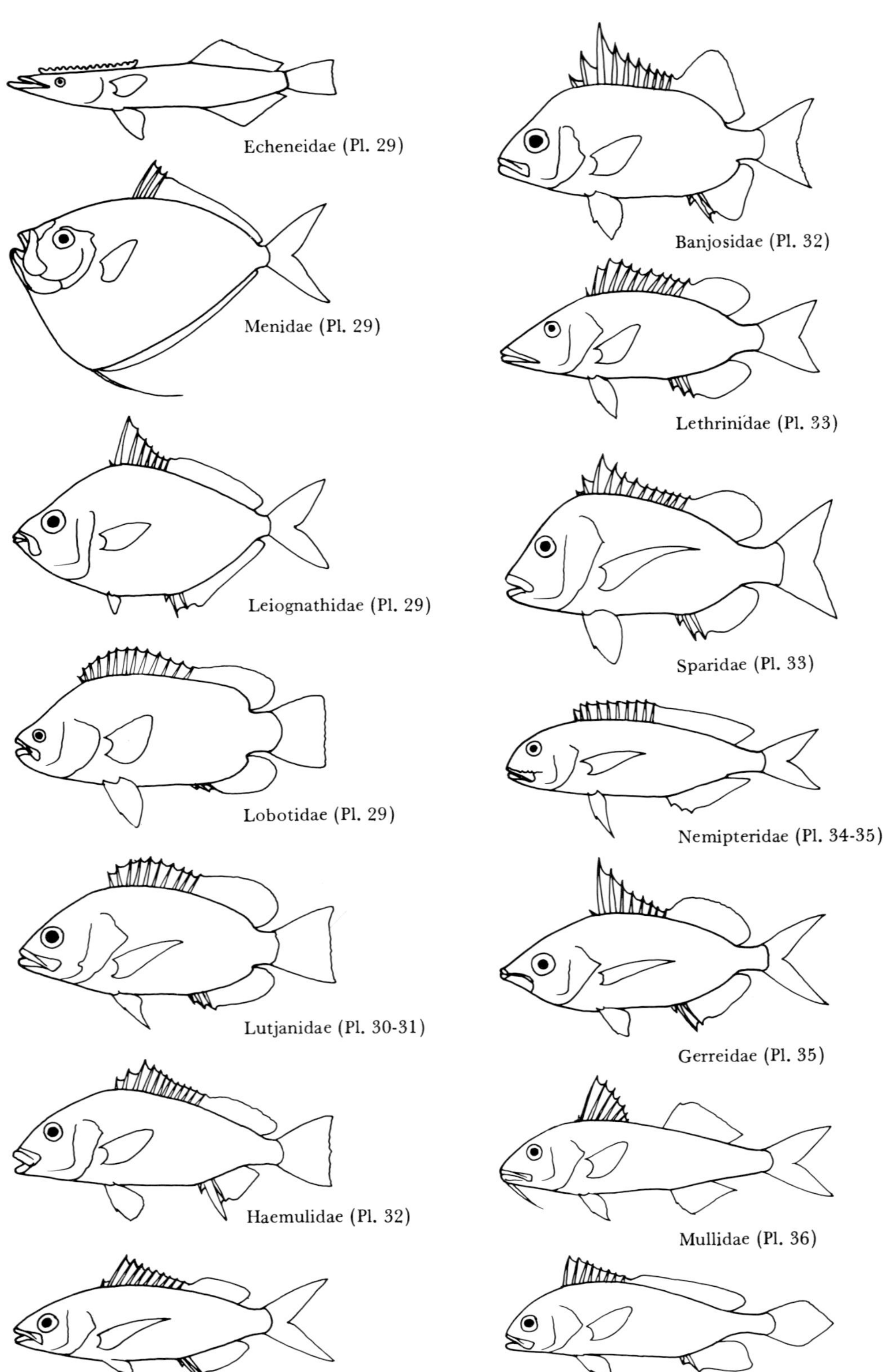

Echeneidae (Pl. 29)

Menidae (Pl. 29)

Leiognathidae (Pl. 29)

Lobotidae (Pl. 29)

Lutjanidae (Pl. 30-31)

Haemulidae (Pl. 32)

Caesionidae (Pl. 32)

Banjosidae (Pl. 32)

Lethrinidae (Pl. 33)

Sparidae (Pl. 33)

Nemipteridae (Pl. 34-35)

Gerreidae (Pl. 35)

Mullidae (Pl. 36)

Sciaenidae (Pl. 37)

14

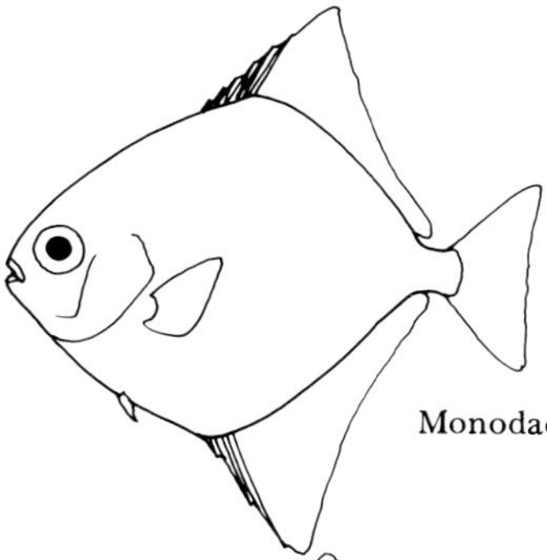

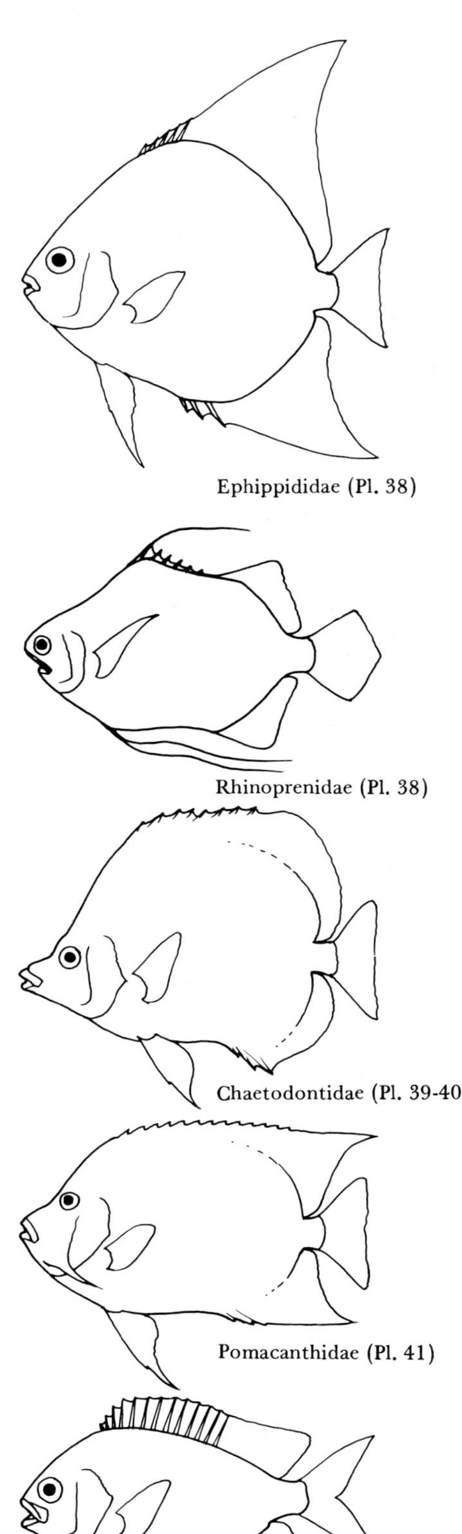

Monodactylidae (Pl. 37)
Leptobramidae (Pl. 37) 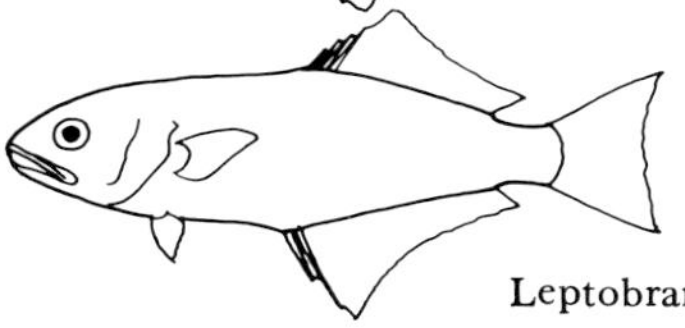
Pempheridae (Pl. 37) 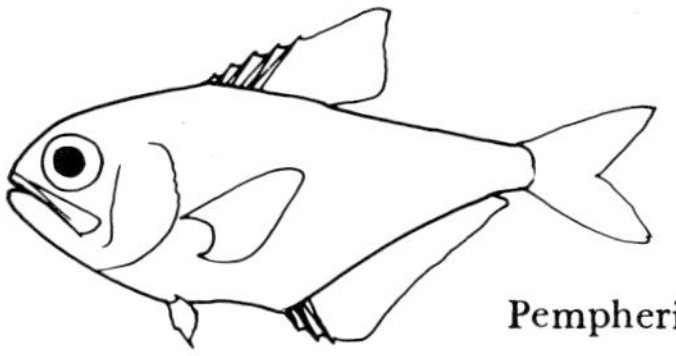
Kyphosidae (Pl. 37) 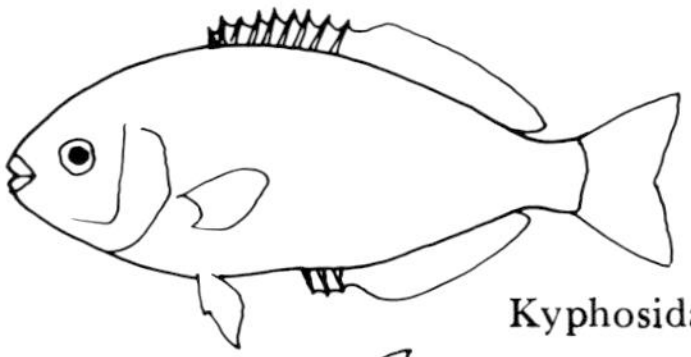
Scorpididae (Pl. 37) 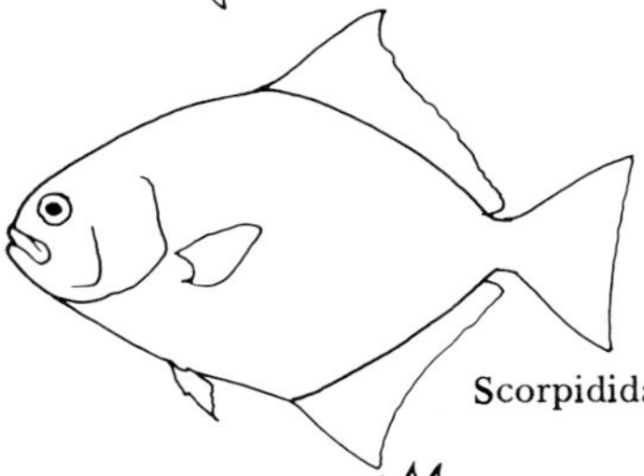
Toxotidae (Pl. 38) 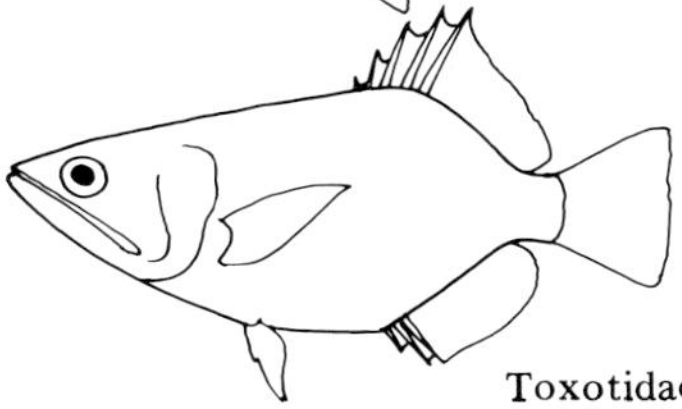
Scatophagidae (Pl. 38) 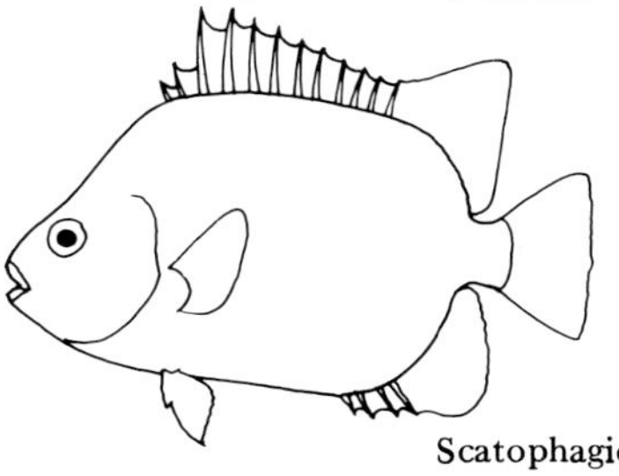
Ephippididae (Pl. 38)
Rhinoprenidae (Pl. 38)
Chaetodontidae (Pl. 39-40)
Pomacanthidae (Pl. 41)
Pomacentridae (Pl. 42-44)

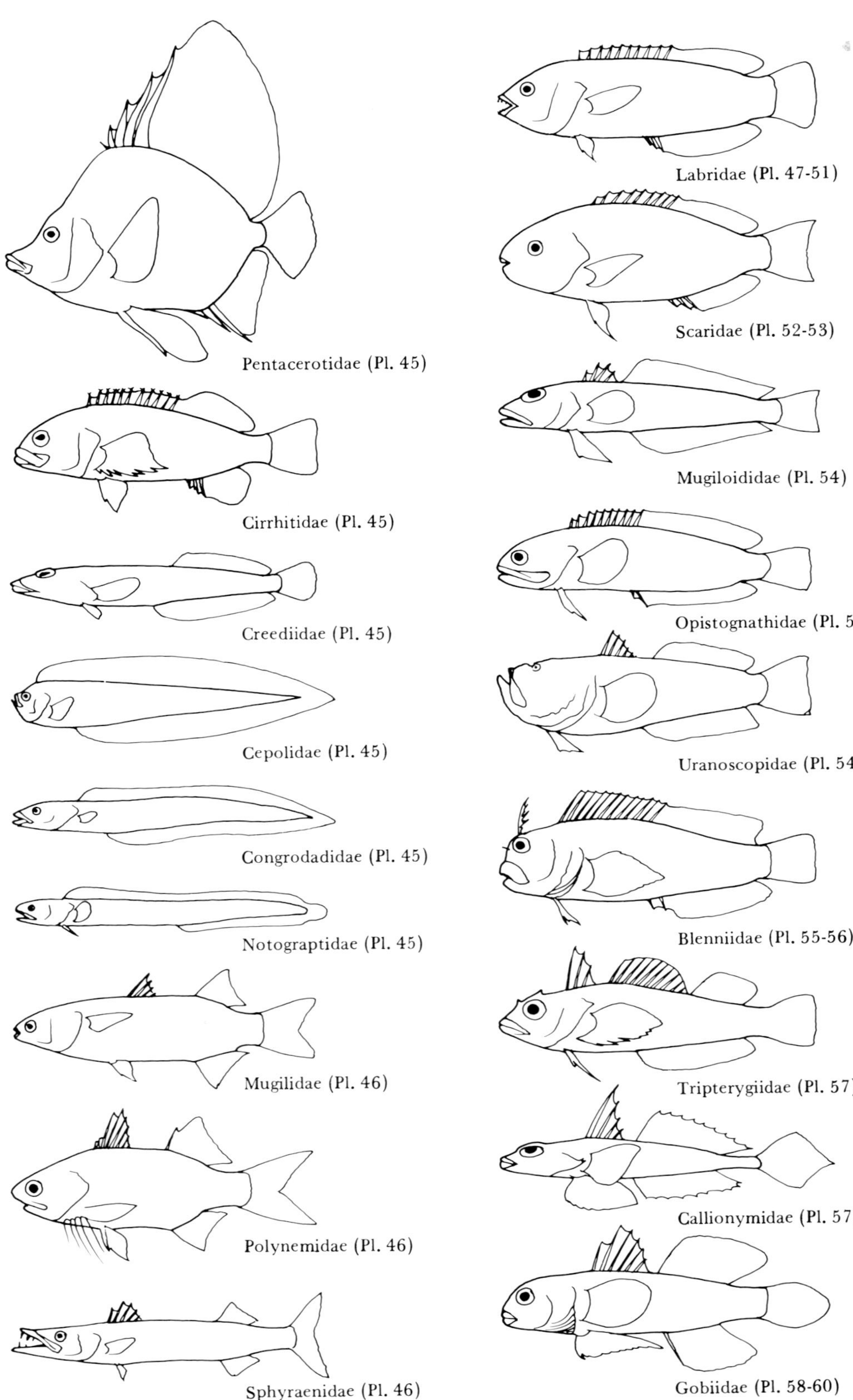

Pentacerotidae (Pl. 45)
Cirrhitidae (Pl. 45)
Creediidae (Pl. 45)
Cepolidae (Pl. 45)
Congrodadidae (Pl. 45)
Notograptidae (Pl. 45)
Mugilidae (Pl. 46)
Polynemidae (Pl. 46)
Sphyraenidae (Pl. 46)
Labridae (Pl. 47-51)
Scaridae (Pl. 52-53)
Mugiloididae (Pl. 54)
Opistognathidae (Pl. 54)
Uranoscopidae (Pl. 54)
Blenniidae (Pl. 55-56)
Tripterygiidae (Pl. 57)
Callionymidae (Pl. 57)
Gobiidae (Pl. 58-60)

16

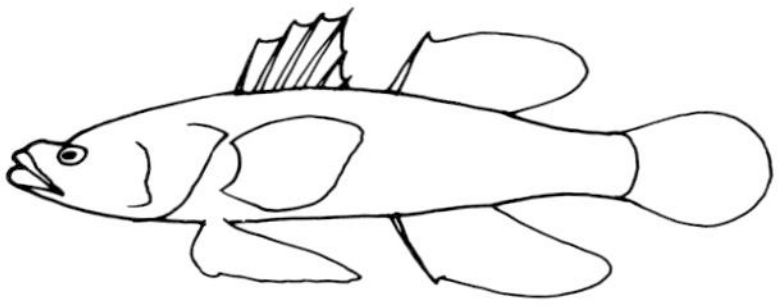

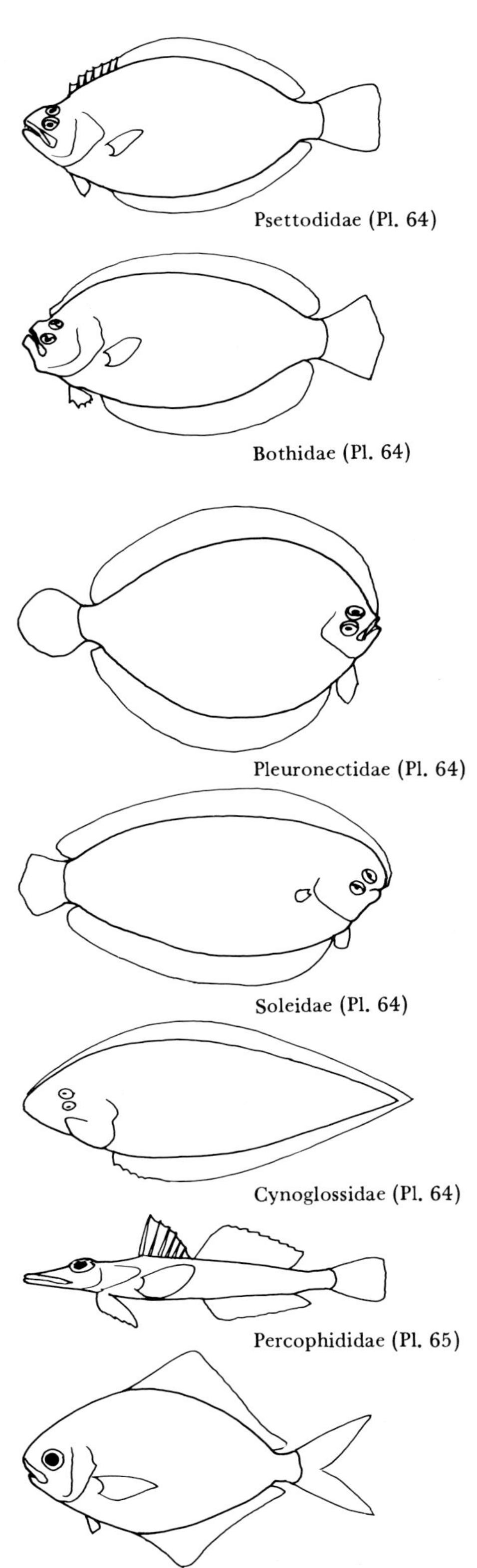

Eleotridae (Pl. 60)
Gobioidae (Pl. 60) 
Zanclidae (Pl. 60) 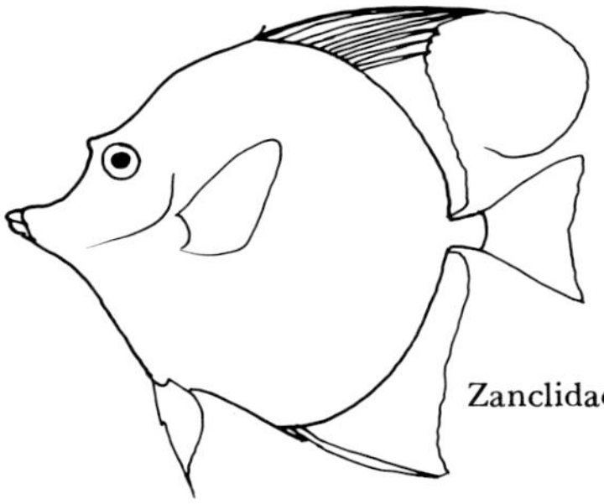
Siganidae (Pl. 60) 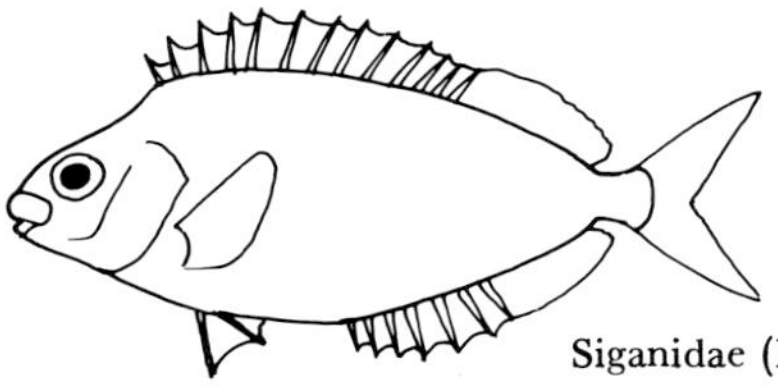
Acanthuridae (Pl. 61) 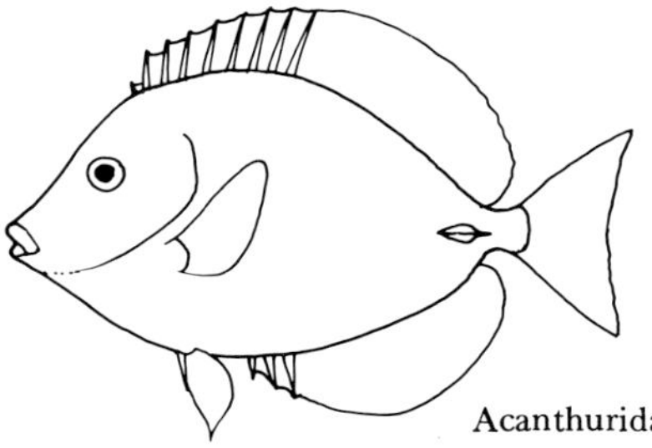
Istiophoridae (Pl. 62) 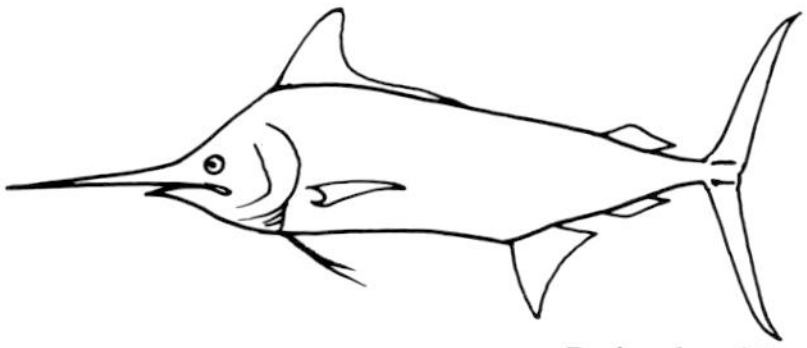
Xiphiidae (Pl. 62) 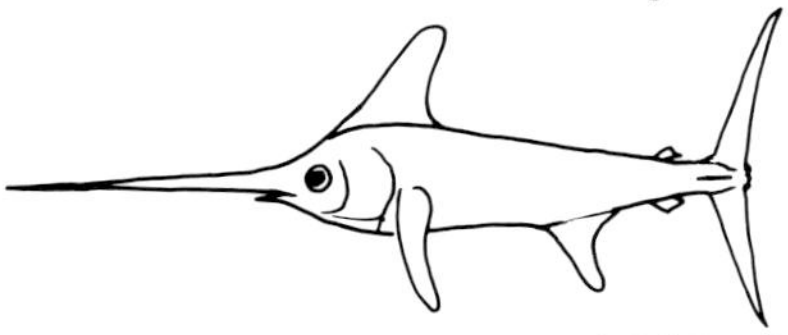
Scombridae (Pl. 62-63) 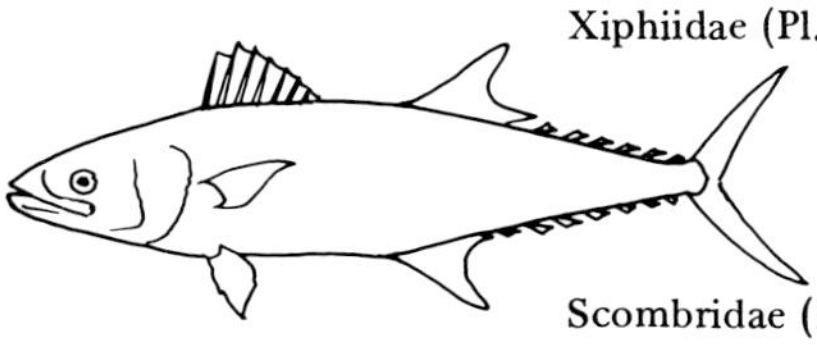
Psettodidae (Pl. 64)
Bothidae (Pl. 64)
Pleuronectidae (Pl. 64)
Soleidae (Pl. 64)
Cynoglossidae (Pl. 64)
Percophididae (Pl. 65)
Centrolophidae (Pl. 65)

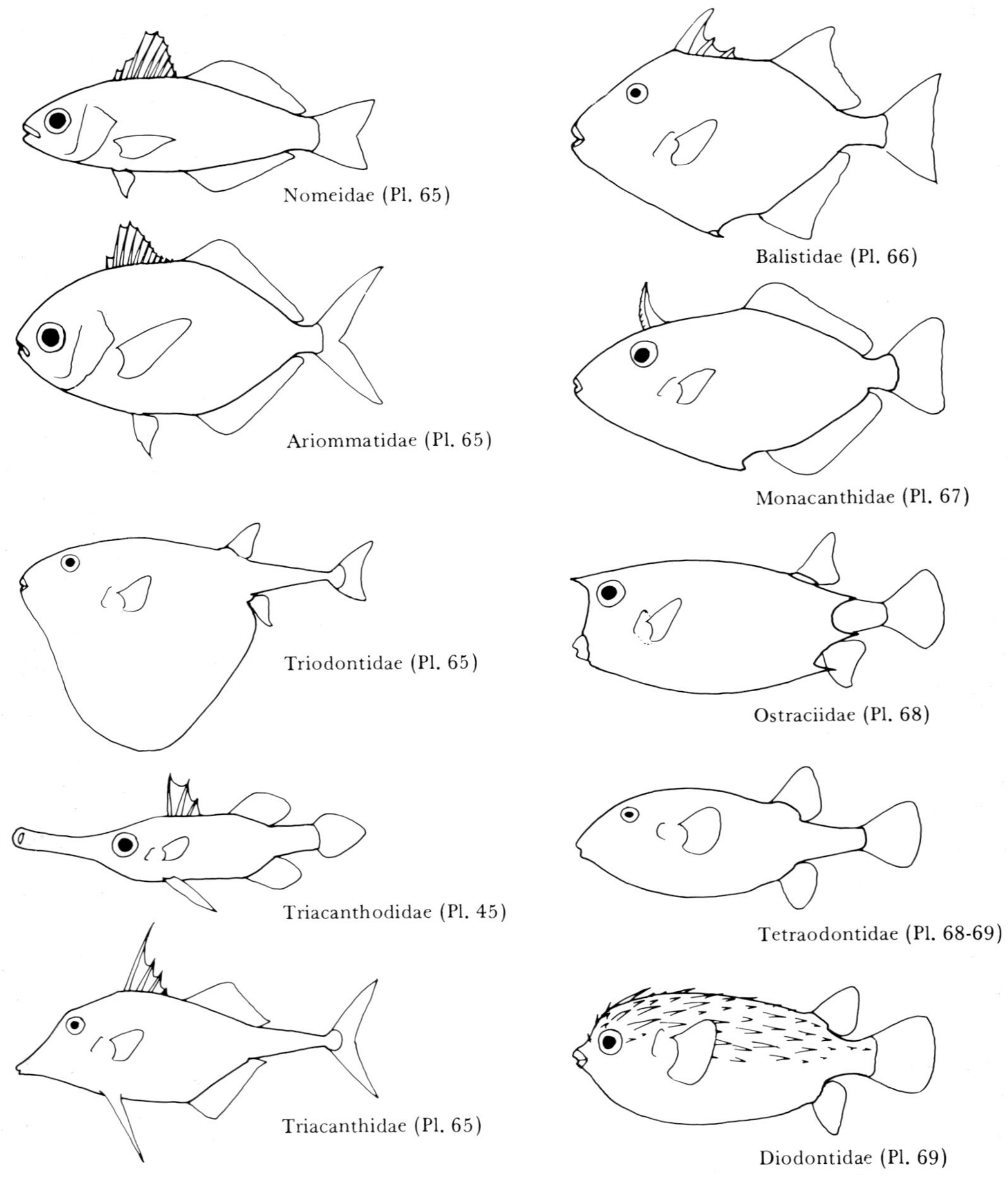

Right: Schools of Bigeye Trevally *(Caranx sexfasciatus)* are common on northern Australian reefs. This ▶ fish grows to nearly a metre in length and is a favourite angling species. (See Plate 27, no. 436 for additional information.) (Photograph P. Kuhn)

1 WHALE SHARK
Rhiniodon typus Smith
Inhabits coastal waters, also occurs well offhore; distinguished by huge size and pattern of white spots; world's largest fish, but harmless plankton-feeder; rarely seen but sightings in the North West Cape area during March-April are a regular occurrence; almost entire coast of W.A.; worldwide temperate and tropical seas; possibly to 18 m; but seldom above 12 m.

2 LEOPARD SHARK
Stegostoma fasciatum (Hermann)
Inhabits coastal waters and offshore areas in the vicinity of coral reefs; may be seen resting on the bottom; distinguished by large tail, dark spots, and ridges on side; also known as Zebra shark; harmless; Port Gregory northwards; Indo-W. Pacific; to 350 cm; 28.575 kg. **★★**

3 TAWNY NURSE SHARK
Nebrius ferrugineus (Lesson)
Inhabits shallow reefs; distinguished by brown colour; equal-sized dorsal fins and moderately long barbels on snout; harmless; Point Quobba northwards; Indo-W. Pacific; to 320 cm; 10.5 kg. **★★**

4 GREY NURSE SHARK
Eugomphodus taurus (Rafinesque)
Inhabits coastal waters; often occurs near the bottom in small schools, distinguished by pair of dorsal and anal fins nearly of equal size, long curved fang-like teeth, and lack of barbels on snout; usually harmless, but will attack if provoked; southern W.A. to Shark Bay; Atlantic and Indo-W. Pacific; to 360 cm; 141.578 kg. **★★**

5 SMALL TOOTH THRESHER SHARK
Alopias pelagicus Nakamura
Inhabits oceanic waters, but occasionally caught near shore; distinguished by very long upper tail lobe (used to stun schools of fish); harmless; Port Hedland northwards; Indo-E. Pacific; to 330 cm.

6 BULLHEAD SHARK
Heterodontus zebra (Gray)
Inhabits flat bottoms on the continental shelf to at least 50 m depth; distinguished by barred pattern and sharp spine at front of both dorsal fins; harmless except dorsal spines can cause painful wound; Broome northwards; mainly W. Pacific; to 122 cm.

7 SHORTFIN MAKO
Isurus oxyrinchus Rafinesque
Inhabits oceanic waters usually well offshore, but sometimes visits coastal areas; distinguished by slender shape, equal-sized tail fin lobes, and slender dagger-like teeth; also known as Blue pointer; dangerous; entire coast of W.A.; worldwide temperate and tropical seas; to 400 cm; 146.4 kg. **★★★**

8 BLUNTNOSE SIXGILL SHARK
Hexanchus griseus (Bonnaterre)
Inhabits coastal waters, also occurs well offshore in deeper waters of the continental shelf; distinguished by absence of second dorsal fin; Onslow northwards; worldwide tropical seas; to 180 cm.

9 SPECKLED CATSHARK
Hemiscyllium trispeculare Richardson
Inhabits shallow coral reefs; distinguished by pale-edged black spot partially surrounded by smaller black spots just behind gill slits; Ningaloo Reef northwards; N. Australia only; to 65 cm.

10 EPAULETTE SHARK
Hemiscyllium ocellatum (Bonaterre)
Inhabits shallow coral reefs; similar to 9, but lacks smaller black spots adjacent to large spot behind gill slits; harmless; Shark Bay northwards; Australia and New Guinea; to 107 cm.

11 BROWN-BANDED CATSHARK
Chiloscyllium punctatum Müller & Henle
Inhabits shallow coral reefs; distinguished by strongly-barred pattern and barbels on snout; harmless; Ningaloo Reef northwards; E. Indian Ocean and W. Pacific; to 104 cm.

12 MARBLED CATSHARK
Atelomycterus macleayi Whitley
Inhabits coastal waters on sand or rocky bottoms; distinguished by small size; slender shape, no barbels on snout, and pattern of black spots and faint broad dark bars; Exmouth Gulf northwards; N. Australia only; to 60 cm.

13 RETICULATED SWELLSHARK
Cephaloscyllium fasciatum Chen
Inhabits deeper waters of the continental shelf; distinguished by rounded, inflatable stomach, blunt snout, narrow eye-slits and pattern of spots and lines; harmless; North West Shelf; N. Australia and S.E. Asia; to 80 cm.

14 BANDED WOBBEGONG
Orectolobus ornatus (de Vis)
Inhabits shallow coastal reefs, frequently on sand-weed bottoms; distinguished by ornate colour pattern and numerous skin flaps on mouth and lower part of head; harmless, but will bite if accidentally trod on; mainly southern W.A. north to the Abrolhos, has been recorded from the north-west coast but needs to be confirmed; Australia only; a similar new species of wobbegong (not shown) with large yellowish blotches and saddles occurs between Cape Leeuwin and Coral Bay; to 300 cm; 73.0 kg.

15 NORTHERN WOBBEGONG
Orectolobus wardi Whitley
Inhabits coastal waters; distinguished by pale-edged dark saddles and bands, frequently has black spots on edge of dorsal fins and tail, skin flaps on head not as well developed as in 14; harmless; Onslow northwards; N. Australia only; to 100 cm.

16 TASSELLED WOBBEGONG
Euchrossorhinus dasypogon (Bleeker)
Inhabits coral reefs; distinguished by numerous branched skin flaps on both chin and side of head (absent on chin in 14 and 15) and very broad, rounded head; harmless; Broome northwards; N. Australia and New Guinea; to 350 cm.

SHARK TEETH

Sharks typically have an outer row of well developed, upright teeth and several inner rows of teeth in various stages of development which are folded downward. Teeth are continuously produced throughout the life of the shark and each row moves forward to replace the next row every few weeks. The teeth are a valuable means of identifying species, particularly among the whalers (17-32). Therefore typical examples from the upper and lower jaw of a number of sharks is included on Plates 25-26.

17 SILVERTIP SHARK
Carcharhinus albimarginatus (Rüppell)
Inhabits offshore coral reefs, usually below 20 m depth on outer edge of reefs; distinguished by white tips on dorsal, tail, and pectoral fins; dangerous; North West Shelf; Indo-E. Pacific; to 300 cm. ★★

18 BIGNOSE SHARK
Carcharhinus altimus (Springer)
Inhabits coastal waters; distinguished by long rounded or bluntly pointed snout when viewed from above, no conspicuous fin markings, and skin ridge between dorsal fins; potentially dangerous; Port Hedland northwards; worldwide tropical seas; to 300 cm. ★★

19 GREY REEF SHARK
Carcharhinus amblyrhynchos (Bleeker)
Inhabits inshore and offshore coral reefs; usually seen adjacent to dropoffs on the outer edge of reefs; distinguished by black margin on tail and lacks skin ridge between dorsal fins; dangerous; Ningaloo Reef northwards; Indo-W. Pacific; to 255 cm. ★★

20 PIGEYE SHARK
Carcharhinus amboinensis (Müller & Henle)
Inhabits coastal waters, sometimes entering estuaries and rivers; a large stout grey shark without distinguishing marks, has a large dorsal fin and lacks a skin ridge between the dorsal fins; dangerous; only W.A. specimen from brackish water in the Prince Regent River (Kimberley coast); Indo-W. Pacific; to 280 cm. ★★

21 BRONZE WHALER
Carcharhinus brachyurus (Günther)
Inhabits coastal waters; often confused with 29, but has narrower upper teeth and no skin ridge between dorsal fins; dangerous; Lancelin northwards; worldwide temperate and tropical seas; to 325 cm; 226.8 kg. ★★★

22 LONG NOSED GREY SHARK
Carcharhinus brevipinna (Müller & Henle)
Inhabits coastal waters, often occurs in schools; distinguished by black tips on most fins, but lacks white margins around black areas as found in 27; potentially dangerous; also known as Smooth-fanged or Inkytail shark; Geographe Bay northwards; Atlantic and Indo-W. Pacific; to 300 cm; 61 kg. ★★★

23 BULL SHARK
Carcharhinus leucas (Valenciennes)
Inhabits coastal waters, enters estuaries and rivers; land-locked freshwater populations occur in some areas outside Australia; a large stocky shark with short blunt snout when viewed from above, broad triangular teeth, and lacks skin ridge between dorsal fins; dangerous; also known as River shark and Estuary whaler; Fremantle northwards; worldwide temperate and tropical seas; to 340 cm. ★★★

24 SILKY SHARK
Carcharhinus falciformes (Bibron)
Inhabits oceanic waters, usually well offshore; a large slender grey shark with a moderately long, rounded snout, short first dorsal fin and elongate tips on anal and second dorsal fins; potentially dangerous; Lancelin northwards; worldwide temperate and tropical seas; to 330 cm; 7.65 kg. ★★★

25 WHITE CHEEK SHARK
Carcharhinus dussumieri Valenciennes
Inhabits coastal waters; see remarks for 31 below; harmless; Port Hedland northwards, N. Indian Ocean and W. Pacific; to 100 cm. ★★★

26 BLACKTIP SHARK
Carcharhinus limbatus (Valenciennes)
Inhabits coastal waters; similar to 22, but more black on fin tips and has black spot on pelvic fin; potentially dangerous, Exmouth Gulf northwards; worldwide temperate and tropical seas; to 255 cm; 15 kg. ★★★

27 BLACKTIP REEF SHARK
Carcharhinus melanopterus (Quoy & Gaimard)
Inhabits reef flats and coral reef lagoons; distinguished from other 'black-tipped' sharks by white margin around black areas, especially noticeable on first dorsal fin; usually not dangerous unless cornered; Ningaloo Reef northwards; Indo-W. Pacific; to 180 cm; 12.6 kg. ★★★

28 OCEANIC WHITETIP SHARK
Carcharhinus longimanus (Poey)
Inhabits oceanic waters, usually well offshore; distinguished by over-sized pectoral fin and broad rounded dorsal fin, both of these fins broadly white-tipped; dangerous; Shark Bay northwards; worldwide temperate and tropical seas; to 396 cm; 56.2 kg. ★★★

29 BLACK WHALER
Carcharhinus obscurus (Le Sueur)
Inhabits coastal waters, also found well offshore; similar to 21, but has wider, more triangular teeth in upper jaw and has a low skin ridge between dorsal fins; dangerous; also known as Dusky shark; Albany northwards; worldwide temperate and tropical seas; to 362 cm; 323.5 kg. ★★★

30 SANDBAR SHARK
Carcharhinus plumbeus (Nardo)
Inhabits coastal waters; distinguished by very tall first dorsal fin that arises above rear base of pectoral fin; dangerous; also known as Sand or Thickskin shark and Northern whaler; entire coast of W.A.; worldwide temperate and tropical seas; to 300 cm, 66 kg. ★★★

31 BLACKSPOT SHARK
Carcharhinus sealei (Pietschmann)
Inhabits coastal waters; similar to 25, but has falcate ($\wedge$) rather than triangular ($\wedge$) first dorsal fin; harmless; Dampier northwards; Indo-W. Pacific; to 95 cm. ★★★

32 SPOT-TAIL SHARK
Carcharhinus sorrah (Valenciennes)
Inhabits coastal waters in the vicinity of coral reefs; distinguished by conspicuous black tips on pectoral and second dorsal fins, and lower lobe of tail; dangerous; Point Quobba northwards; Indo-W. Pacific; to 160 cm. ★★★

> ### SHARK ATTACK!
> The sharks illustrated in Plate 2 are members of the family Carcharhinidae, commonly known as whalers. Although many of the species have never been implicated as far as attacks on humans are concerned, the family contains several which have a bad reputation. All should be handled with respect when removing hooks and none should be deliberately provoked by spearing fishes or offering food when diving in their company.

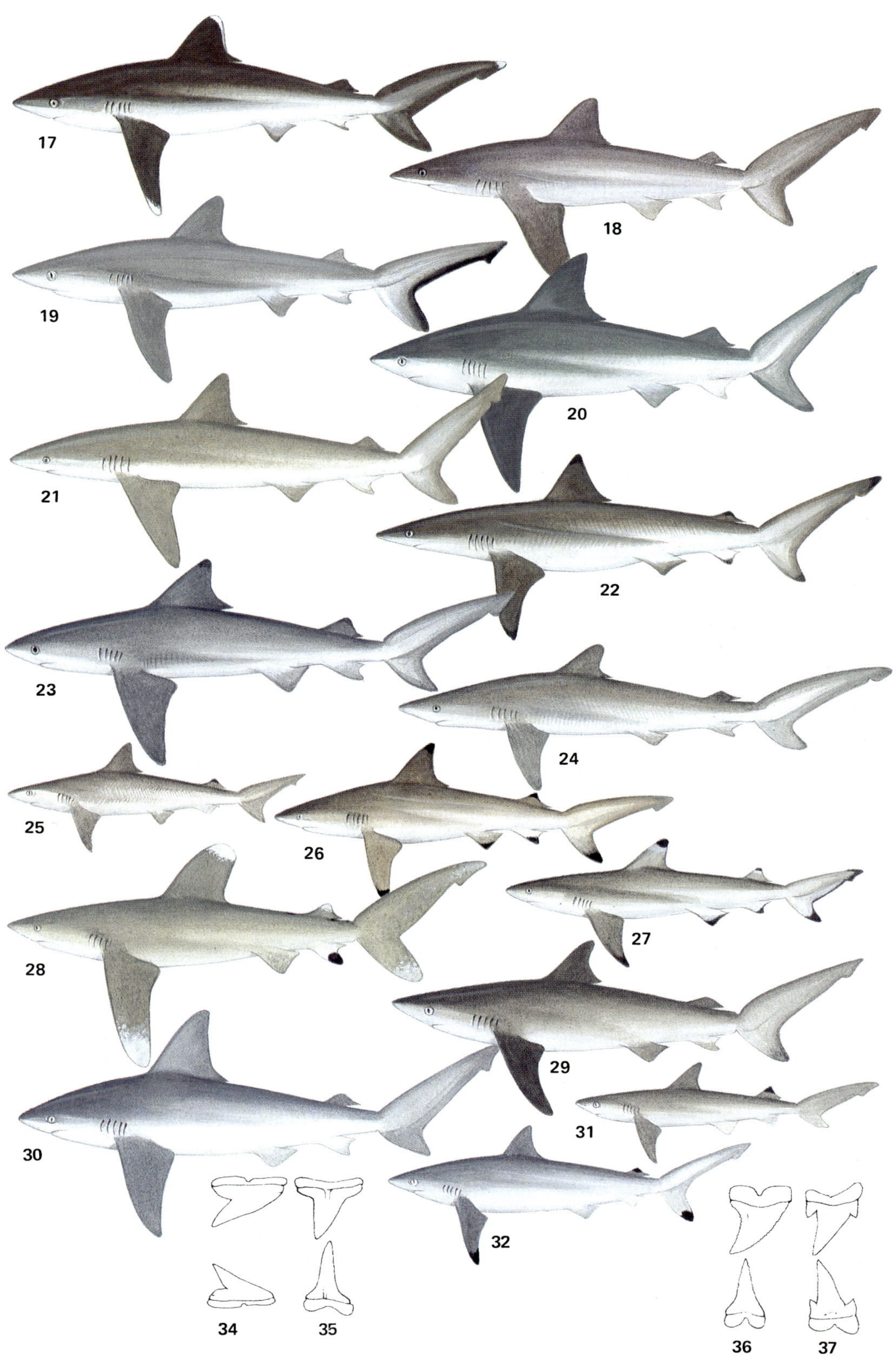

33 TIGER SHARK
Galeocerdo cuvier (Peron & Le Sueur)
Inhabits deeper offshore areas, frequently near reefs; distinguished by blunt (when viewed from above) snout, stripes on side (faint or absent in large adults), keel on side of tail base, and strongly curved teeth; very dangerous; entire coast of W.A., but rare south of Jurien Bay; worldwide temperate and tropical seas; to 650 cm; 521.0 kg. ★★★

34 SLITEYE SHARK
Loxodon macrorhinus Müller & Henle
Inhabits continental shelf waters between 7-80 m depth; distinguished by notch or slit on rear edge of eye socket, long slender snout, and large eye; harmless; North West Cape northwards; Indo-W. Pacific; to 91 cm. ★★★

35 LEMON SHARK
Negaprion acutidens (Rüppell)
Inhabits inshore waters, in bays, estuaries and in coral reef lagoons; distinguished by yellow-brown colour, short snout and stocky body with 2 dorsal fins about equal sized; generally harmless to divers, but potentially dangerous; Rottnest Island northwards, but rare south of Abrolhos; Indo-W. Pacific; to 335 cm; 90.718 kg. ★★★

36 BLUE SHARK
Prionace glauca (Linnaeus)
Inhabits surface waters, usually well offshore; similar to Mako (7) at least in colour, but has smaller gill slits; longer pectoral fins and serrated teeth; also called Blue whaler; dangerous; entire coast of W.A.; worldwide temperate and tropical seas; to 380 cm; 140 kg. ★★★

37 WHITETIP SHARK
Triaenodon obesus (Rüppell)
Inhabits coral reefs, frequently seen resting on the bottom in caves or in the open; distinguished by slender shape and white tips on first dorsal and caudal fin; usually harmless, but has attacked humans; Point Quobba northwards; Indo-E. Pacific; to 215 cm. ★★

38 MILK SHARK
Rhizoprionodon acutus (Rüppell)
Inhabits coastal bays and off sandy beaches, also offshore areas to 200 m depth; similar to 34, but lacks notch on rear edge of eye socket; harmless Broome northwards; E. Atlantic and Indo-W. Pacific; to 178 cm. ★★

39 SCALLOPED HAMMERHEAD
Sphyrna lewini (Griffith & Smith)
Inhabits coastal waters and also encountered well offshore, frequently near the surface; distinguished from other W.A. hammerheads by an indentation in the middle of the front edge of the head; dangerous; Geographe Bay northwards, but rare south of Abrolhos; worldwide temperate and tropical seas; to 420 cm; 76 kg. ★★★

40 WINGHEAD SHARK
Eusphyra blochii (Cuvier)
Inhabits shallow coastal waters; distinguished by huge wing-shaped head (nearly ½ length of body); considered harmless; Port Hedland northwards; N. Indian Ocean and Indo-Australian Archipelago; to 152 cm. ★★

41 GREAT HAMMERHEAD
Sphyrna mokarran (Rüppell)
Inhabits coastal waters and also found well offshore; distinguished by very flat front edge of head and tall, sail-like first dorsal fin with pointed tip, otherwise body shape similar to 39; dangerous; Abrolhos northwards; worldwide temperate and tropical seas; to 610 cm. ★★★

42 SMOOTH HAMMERHEAD
Sphyrna zygaena (Linnaeus)
Inhabits coastal waters and also found well offshore; distinguished from other W.A. hammerheads by smooth front edge of head; dangerous; entire coast of W.A.; worldwide temperate and tropical seas; to 400 cm; 96 kg. ★★★

43 GREEN SAWFISH
Pristis zijsron Bleeker
Inhabits mud bottoms, entering estuaries; distinguished by relatively long saw-like snout, similar to 44, but has teeth extending along entire edge of snout; not dangerous unless cornered; Broome northwards: Indo-W. Pacific; to 600 cm; 38 kg. ★★★

44 NARROW SAWFISH
Pristis cuspidatus Latham
Inhabits mud bottoms, entering estuaries; similar in appearance to 43, but has very long, narrow snout with teeth absent on rear part; Derby northwards; N. Indian Ocean and Indo-Australian Archipelago; to 350 cm. ★★★

45 WIDE SAWFISH
Pristis microdon Latham
Inhabits mud bottoms of bays and estuaries, also enters large rivers and goes well upstream; similar to 43, but has shorter, broader snout. Broome northwards; Indo-W. Pacific; to 460 cm, 1.5 kg. ★★★

46 SHARK RAY
Rhina ancylostoma (Bloch & Schneider)
Inhabits coastal waters, on mud or sand bottoms; distinguished from 47 by rounded head and granular patches or ridges above each eye and on middle of forehead; Exmouth Gulf northwards; Indo-W. Pacific; to 260 cm; 14.305 kg. ★★

47 WHITE-SPOTTED SHOVELNOSE RAY
Rhynchobatus djiddensis (Forsskål)
Inhabits sandy areas, sometimes seen resting motionless; distinguished from 46 by pointed head and black spot above pectoral fin base; sometimes incorrectly referred to as Shovelnose shark; Fremantle northwards; Indo-W. Pacific; to 300 cm; 75 kg. ★★★

48 SPOTTED SHOVELNOSE RAY
Aptychotrema sp.
Inhabits coastal waters on sand bottoms; similar to 49, but has broad brown margin around white spots; Port Hedland northwards; entire distribution unknown; to 120 cm. ★★★

49 GOLDEN-EYED SHOVELNOSE RAY
Rhinobatos sp.
Inhabits coastal waters; similar to 48, but lack spots and snout tip is rounded; Port Hedland northwards; enitre distribution unknown; to 100 cm. ★★★

50 ORNATE ANGEL SHARK
Squatina tergocellata McCulloch
Inhabits deep trawling grounds, a ray-like shark with distinctive shape, has greatly enlarged pectoral fins that are not entirely fused to the head and body as in rays, also has bilobed tail and 2 small dorsal fins; entire coast of W.A.; Australia only; to 51 cm. ★★

51 YELLOW SHOVELNOSE-RAY
Aptychotremata sp.
Inhabits sand bottoms; similar to 48 on previous page, but lacks spots and snout is blunter; continental shelf of north-western Australia; to at least 65 cm. ★★

52 BROWN STINGAREE
Urolophus westraliensis Last & Gomon
Inhabits sand bottoms in depths of 150-210 m, distinguished by sharply pointed snout tip, at least one serrated spine on tail, similar to 53 and 68 (next page) but is plain brown without markings; outer continental shelf of north-western Australia; to at least 36 cm. ★★

53 BLOTCHED STINGAREE
Urolophus mitosis Last & Gomon
Inhabits sand bottoms to depths of 200 m; similar to 52 and 68 (next page) but has pale elongate blotches and lines that surround clusters of dark spots; outer continental shelf of north-western Australia; to at least 29 cm. ★★

54 ORNATE NUMBFISH
Narcine westraliensis McKay
Inhabits sand bottoms; distinguished by flattened 'tadpole' shape, large round head, and 2 small dorsal fins, colour ranges from plain to strongly spotted; capable of producing mild electrical shock; Shark Bay northwards; N.W. Australia only; to 28 cm.

55 NUMBFISH
Hypnos monopterygium (Shaw & Nodder)
Inhabits sand-weed areas; distinguished by round body with smaller rounded pelvic lobe at rear which bears the tail and 2 small dorsal fins, colour varies from light brown to blackish; can produce a strong electrical shock if handled or accidently trod on; most of W.A. coast; Australia only; to 69 cm; 2.702 kg.

56 EYED SKATE
Raja sp.
Inhabits sand bottoms; distinguished by pointed snout, bilobed pelvic fins, single row of thorns or small spines down middle of tail, and 2 small dorsal fins near end of tail, also has fragmented ocellus-type markings on each side of back; continental shelf of north-western Australia; to at least 20 cm width.

57 ROUND SKATE
Irolita waitii (McCulloch)
Inhabits sand bottoms; distinguished by round shape, bilobed pelvic fins, short spines or thorns in 1 or more rows on tail, and 2 small dorsal fins near end of tail, also has blue-grey spots or blotches scattered on back; entire coast of W.A.; Australia only; to 45 cm.

58 BROWN STINGRAY
Dasyatis annotatus Last
Inhabits sand bottoms in 40-65 m depth; distinguished by kite-shape and 2 long serrated spines on tapering tail, similar to 59 and 60, but is plain grey-brown without markings and has a more pointed snout; Timor and Arafura seas off northern Australia; to at least 24 cm width. ★★

59 BROWN RETICULATED STINGRAY
Dasyatis leylandi Last
Inhabits sand bottoms; similar to 58 and 60, but has numerous pale blotches and reticulated pattern of brown lines, Port Hedland northwards; entire distribution unknown; to at least 25 cm width. ★★

60 BLACK-SPOTTED STINGRAY
Dasyatis sp.
Inhabits sand bottoms; similar to 59, but is pinkish or light brown with pepper-like spotting on back; Port Hedland northwards; entire distribution unknown; to at least 28 cm width. ★★

61 BLUE-SPOTTED STINGRAY
Dasyatis kuhlii (Müller & Henle)
Inhabits sand bottoms, frequently in the vicinity of coral reefs; sometimes buries itself and only the eyes protrude above the sand; distinguished by blue spots and frequently there are scattered black spots on the disc; Indo-W. Pacific; maximum width about 45 cm, length to 70 cm; 7.5 kg.

SHARK FACTS

Sharks and their cousins, the rays, are very specialised fishes that represent a primitive stage of evolutionary development. The sharks represent an extremely ancient lineage. They were known in Devonian seas, over 350 million years ago. Many of the present genera of sharks, skates, and rays date back more than 100 million years. Approximately 350 species of sharks are currently known. Only a small number of these are considered dangerous. Sharks represent an extremely diverse assemblage, occurring in all seas, and in a variety of depths ranging from shallow intertidal pools to deep oceanic trenches, miles below the surface.

RAYS AND RELATIVES

Rays and their relatives are classified in the Order Rajiformes. Both sharks and rays are characterised by a cartilagenous skeleton. Other typical features of rays include a greatly flattened body that is often disc-shaped, and the presence of five, ventrally located gill openings. Most species bear their young alive except the Rajidae (56-57), which deposit egg cases. Many rays have venomous spikes or spines on the tail base that are capable of inflicting painful wounds. The members of the family Torpedinidae (54-55), possess powerful electric organs situated in the head region. Rays dwell in a variety of habitats, ranging from oceanic depths to shallow reefs, estuaries, and even freshwater streams. They range in size from about 30-40 cm disc-width (some skates) to more than 4 m (manta rays).

62 BLACK STINGRAY
Dasyatis thetidis Waite
Inhabits coastal waters off beaches and over sand or mud bottoms to at least 300 m depth; distinguished by blue-grey to blackish colour and short tubercles on top of head and over middle of back; has a pair of spines on tail that can inflict serious wounds; Jurien Bay northwards; Australia, New Zealand, and South Africa; to 180 cm disc width, 400 cm total length; 65.499 kg. ★★

63 COWTAIL STINGRAY
Dasyatis sephen (Forsskål)
Inhabits flat sand or mud bottoms near shore, also common in brackish mangrove estuaries and in the lower reaches of rivers; distinguished by grey-brown to blackish colour and broad flap of skin on lower edge of tail, large specimens may have tubercles on back similar to 62; has 2 dangerous spines on tail; Port Hedland northwards; Indo-W. Pacific to 180 cm total length. ★★

64 BLACK-BLOTCHED STINGRAY
Taeniura melanospila (Bleeker)
Inhabits sandy bottoms in the vicinity of coral reefs; distinguished by round shape and dense pattern of black spots; has pair of dangerous spines on tail; Ningaloo Reef northwards; Indo-W. Pacific; to 165 cm disc width and 300 cm total length. ★★

65 BLUE-SPOTTED FANTAIL STINGRAY
Taeniura lymma (Forsskål)
Inhabits flat sand bottoms in the vicinity of coral reefs; distinguished by kite shape and pattern of bright blue spots; has 1-2 dangerous spines on tail; Ningaloo Reef northwards; Indo-W. Pacific; to 95 cm disc width; 240 cm total length; 1.499 kg.

66 COACHWHIP STINGRAY
Himantura uarnak (Forsskål)
Inhabits sandy beaches, sand flats near reefs, or shallow mangrove estuaries; distinguished by very long, whip-like tail, 2 colour varieties are known, one with small black spots and another with a dense network of blotches; has pair of dangerous spines on tail; Dampier northwards; Indo-W. Pacific; to 200 cm disc width; 600 cm total length (mainly tail); 4.5 kg. ★★

67 MANTA RAY
Manta birostris (Donndorff)
Inhabits coastal waters and the vicinity of offshore reefs; distinguished by large size, pair of protruding flaps at front of head, and short tail; one of the largest of all fishes, it is a harmless plankton feeder well known for its ability to make spectacular leaps above the water surface; Rottnest Island northwards; worldwide circumtropical; to 600 cm disc width and over 2 tons in weight. ★★★

68 PATCHWORK STINGAREE
Urolophus sp.
Inhabits coastal waters on flat sand bottoms; similar to 52 and 53 on previous page, but has ornate pattern of white spots and reticulated white and brown lines; North West Cape northwards; N. Australia, but possibly more widespread; to 40 cm disc width. ★★

69 RAT-TAILED RAY
Gymnura australis (Ramsay & Ogilby)
Inhabits shallow coastal waters; distinguished by broad triangular "wings" and very short rat-like tail; also known as Butterfly ray; Dampier northwards; Australia only; to 450 cm disc width.

70 BARBLESS EAGLE RAY
Aetomyleus nichofii (Bloch & Schneider)
Inhabits coastal waters in the vicinity of reefs; distinguished by bulging head and protruding snout, similar to 70, but lacks white spots and has pale blue cross bands on back; Ningaloo Reef northwards; Indo-W. Pacific; to 350 cm disc width. ★★

71 SPOTTED EAGLE RAY
Aetobatus narinari (Euphrason)
Inhabits coastal waters in the vicinity of reefs; similar to 70, but has white spots on back and 2-6 barbed spines on base of tail; Ningaloo Reef northwards; worldwide circumtropical; to 350 cm disc width; 8.75 kg. ★★

72 GHOST SHARK
Hydrolagus sp.
Inhabits deeper offshore trawling grounds; distinguished by 'rodent-like' head with small mouth below eye, large pectoral fins, and long tapering tail; Port Hedland northwards; entire distribution unknown; to 30 cm.

> ### BEWARE OF SPINES!
>
> Many of the rays illustrated on Plates 4 and 5 are characterised by one or more venomous spines on the tail. Stings inflicted by these spines are extremely painful and fatalities have occurred when either heart, abdomen, or lungs were badly perforated. Caution should be exercised when wading on sandy bottoms. It is advisable to use a walking stick to probe just ahead or at least walk with a shuffling gait rather than in normal strides. This sort of movement will often prevent treading directly on the back of a partially buried ray. If a ray is stepped on it has the ability to thrust its tail upward and forward, impaling the victim with remarkable speed. Pain is immediate and intense, and may persist for several days. Immersion in hot water (about 50°C) for 30-90 minutes may dramatically relieve the pain as the venom is a protein which is heat labile. Medical assistance should always be obtained as the wound may be far more serious than it appears.

73 GIANT HERRING
Elops hawaiiensis Regan
Inhabits coastal waters and mangrove areas; distinguished by slender body and relatively large mouth; Albany northwards; Indo-W. Pacific; to 120 cm. ★

74 BONEFISH
Albula vulpes (Linnaeus)
Inhabits estuaries and mudflats; distinguished by protruding snout; Shark Bay northwards; Indo-W. Pacific; to 110 cm. ★

75 MILKFISH
Chanos chanos (Forsskål)
Inhabits coastal waters near reefs; distinguished by small mouth and scissor-like tail; Shark Bay northwards; Indo-W. Pacific; to 120 cm; 10.6 kg. ★

76 OXEYE HERRING
Megalops cyprinoides Broussonet
Inhabits coastal waters and mangrove areas; distinguished by large eye and mouth, and filament at end of dorsal fin; Onslow northwards; Indo-W. Pacific; to 150 cm; 2.3 kg. ★

77 WOLF HERRING
Chirocentrus dorab Forsskål
Inhabits coastal waters; distinguished by large fangs; Broome northwards; Indo-W. Pacific; to 140 cm. ★

78 SMOOTH-BELLY SARDINE
Amblygaster leiogaster (Valenciennes)
Inhabits coastal waters in large schools; similar to 80, but pelvic and anal fins farther apart; Ningaloo Reef northwards; Indo-W. Pacific; to 25 cm. ★★

79 NORTHERN PILCHARD
Amblygaster sirm (Walbaum)
Inhabits coastal waters; distinguished by row of spots on side; Onslow northwards; Indo-W. Pacific; to 25 cm. ★★

80 SLENDER SARDINE
Dussumieria elopsoides Bleeker
Inhabits coastal waters; similar to 78, but pelvic and anal fins closer together; Broome northwards; Indo-W. Pacific; to 23 cm. ★★

81 BLUESTRIPE HERRING
Herklotsichthys quadrimaculatus (Rüppell)
Inhabits coastal waters; distinguished from all other plain coloured herrings on this page by a pair of fleshy outgrowths on margin of gill cover; Onslow northwards; Indo-W. Pacific; to 16 cm. ★★

82 KONINGSBERGER'S HERRING
Herklotsichthys koningsbergeri (Weber & de Beaufort)
Inhabits beaches and inlets; distinguished by double row of spots on side; Shark Bay northwards; N.W. Australia Bay; to 15 cm. ★★

83 GIZZARD SHAD
Anodontostoma chacunda (Hamilton)
Inhabits coastal waters and mangrove areas; similar to 86, but lacks filament at rear of dorsal fin; Broome northwards; Indo-Australian Archipelago and N. Indian Ocean; to 20 cm. ★★

84 BANDED ILISHA
Ilisha striatula Wongratana
Inhabits coastal waters; similar to 85, but has faint stripe along middle of side (not shown) and lacks dark spot behind gill cover; Broome northwards; Indian Ocean; to 22 cm. ★★

85 DITCHELEE
Pellona ditchela Valenciennes
Inhabits coastal bays and estuaries; similar to 84, but lacks stripe on sides and has dark spot behind gill cover; Onslow northwards; mainly Indian Ocean and Indo-Australian Archipelago; to 18 cm. ★★

86 HAIRBACK HERRING
Nematalosa come (Richardson)
Inhabits coastal bays and estuaries; similar to 83 but has filament at end of dorsal fin; Kalbarri northwards; mainly Indo-Australian Archipelago; to 23 cm. ★★

87 SLENDER SPRAT
Spratelloides gracilis (Temminck & Schlegel)
Inhabits coastal waters; distinguished by slender shape and silvery stripe on sides; Abrolhos northwards; Indo-W. Pacific; to 11 cm. ★★

88 BLUE SPRAT
Spratelloides robustus Ogilby
Inhabits coastal waters and estuaries; distinguished by bluish back and lack of silver stripe on sides; Geographe Bay northwards; Australia only; to 9 cm; 0.110 kg. ★★

89 GOLDSTRIPE SARDINE
Sardinella gibbosa (Bleeker)
Inhabits coastal waters; distinguished by thin gold-coloured stripe on sides; Shark Bay northwards; Indo-W. Pacific; to 19 cm. ★★★

90 BAREBACK ANCHOVY
Papuengraulis micropinna Munro
Inhabits coastal bays and estuaries; distinguished by thread-like dorsal fin; Derby northwards; Australia and New Guinea; to 15 cm. ★★

91 INDIAN ANCHOVY
Stolephorus indicus (van Hasselt)
Inhabits coastal waters; distinguished by rounded snout and broad silvery stripe on sides; Broome northwards; Indo-W. Pacific; to 16 cm. ★★

92 LONGFIN ANCHOVY
Setipinna tenuifilis Valenciennes
Inhabits coastal waters; similar to 90, but has normal dorsal fin; Derby northwards; Indo-W. Pacific; to 20 cm. ★★

93 HAMILTON'S ANCHOVY
Thryssa hamiltonii (Gray)
Inhabits estuaries and mudflats; distinguished by rounded snout, large mouth, and spot behind gill cover, *Thryssa setirostris* (not shown) similar, but with extermely long posterior extension of upper jaw; Exmouth Gulf northwards; Indo-Australian Archipelago and Indian Ocean; to 25 cm.

HERRINGS AND ALLIES

The herrings and other species featured on this plate are primitive bony fishes which are generally silvery in colour, lack spiny fin rays and have a single dorsal fin. Most are inhabitants of tropical and subtropical seas, although they are frequently found in estuaries or in the lower reaches of freshwater streams. Herrings (78-89) and anchovies (90-93) are the basis of valuable commercial fisheries in many parts of the world.

94 STARRY EEL
Echidna nebulosa (Ahl)
Inhabits shallow coral reefs; distinguished by whitish body with 2 longitudinal rows of darkish pale-centred blotches and lacks sharp fangs; also known as Clouded reef-eel; Ningaloo Reef northwards; Indo-E. Pacific; to 70 cm.

95 GIRDLED REEF EEL
Echidna polyzona (Richardson)
Inhabits shallow coral reefs, often exposed to surge; distinguished by alternating light and dark bars of approximately equal width and lacks sharp fangs; North West Shelf; Indo-W. Pacific; to 60 cm.

96 ZEBRA EEL
Echidna zebra (Shaw)
Inhabits shallow coral reefs, often exposed to surge; distinguished by narrow pale bands encircling head and body, and lacks sharp fangs; Ningaloo Reef northwards; Indo-E. Pacific; to 150 cm.

97 LATTICE-TAIL MORAY
Gymnothorax buroensis (Bleeker)
Inhabits offshore coral reefs; distinguished by brown colour on front of body and blackish colour on posterior part with pale spotting; North West Shelf; Indo-E. Pacific; to 31 cm.

98 SPOTTED MORAY
Gymnothorax eurostus (Abbot)
Inhabits coral reef crevices; distinguished by numerous small yellowish spots becoming larger on rear part of body, also dark spots or blotches evident mainly on front half; Shark Bay northwards; Indo-E. Pacific; to 40 cm. 0.23 kg.

99 SIEVE-PATTERNED MORAY
Gymnothorax criboris Whitley
Inhabits coral reef crevices; distinguished by several dark spots behind eye, network of fine interconnected lines on front half of body, and network of darker brown surrounding pale blotches on posterior half; also known as Brown-flecked reef eel; Dampier northwards; N. Australia only; to 75 cm.

100 BLACK-BLOTCHED MORAY
Gymnothorax favagineus Bloch & Schneider
Inhabits coral reef crevices; distinguished by bold spot pattern; one of the largest of moray eels; but usually harmless unless provoked, its sharp fangs can cause serious injury; also known as Tesselated moray and Giraffe eel; Barrow Island northwards; Indo-W. Pacific; to 300 cm; 2.041 kg.

101 YELLOW-EDGED MORAY
Gymnothorax flavimarginatus (Rüppell)
Inhabits coral reef crevices; generally yellow-brown in colour with fine dark spotting on head and body and black patch at gill opening, juveniles are dark brown with fine yellow-green margin on dorsal and anal fins; also known as Leopard eel; Ningaloo Reef northwards; Indo-E. Pacific; to 50 cm.

102 FIMBRIATED MORAY
Gymnothorax fimbriatus (Bennett)
Inhabits coral reef crevices; distinguished by tan or light brown colour with loose network of branched dark bands; Onslow northwards, Indo-W. Pacific; to 80 cm.

103 GIANT MORAY
Gymnothorax javanicus (Bleeker)
Inhabits offshore coral reef; distinguished by yellow-brown head with small dark spots and large dark patch at gill opening, adults have leopard-like spotting on body; a large eel that can be dangerous if provoked, several attacks have been reported; North West Shelf; Indo-W. Pacific; to 250 cm.

104 PEARLY MORAY
Gymnothorax margaritophorus (Bleeker)
Inhabits coral reef crevices; distinguished by series of dark blotches just behind eye, pale chin and breast, and 'lattice' pattern on rear part of body; Ningaloo Reef northwards; Indo-W. Pacific; to 40 cm.

105 MOTTLED MORAY
Gymnothorax undulatus (Lacepède)
Inhabits coral reef crevices; distinguished by 'chain-link' pattern of narrow pale bands, juvenile with diffuse vertical bars most noticeable towards tail; Abrolhos northwards; Indo-E. Pacific; to 150 cm; 2.42 kg.

106 BARTAIL MORAY
Gymnothorax zonipectus Seale
Inhabits coral reef crevices; distinguished by white spots on upper and lower jaw, dark spots on body, and distinct dark bars on rear portion of dorsal and anal fins; Ningaloo northwards; Indo-W. Pacific; to 32 cm.

107 PAINTED MORAY
Siderea picta (Ahl)
Inhabits shallow reef flats and tide pools, sometimes seen entirely out of water at low tide; distinguished by white colouration with numerous small dark spots; also known as Peppered moray; Ningaloo Reef northwards; Indo-E. Pacific; to 68 cm.

108 FRECKLED MORAY
Siderea thrysoidea (Richardson)
Inhabits shallow coral reefs; distinguished by light brown or tan coloured body (with faint mottlings), white to bluish snout, and silvery eyes; Point Quobba northwards; Indo-W. Pacific; to 35 cm.

MORAY EELS

The eels featured on this plate are all members of the family Muraenidae, commonly known as morays. Most are equipped with needle-sharp teeth which has given them a largely undeserved reputation of being dangerous. While it is true that some larger eels, for example no. 103, have attacked humans, in most cases the eel had been provoked in some manner. Large eels should definitely not be teased with offerings of dead or struggling fish either hand held or on the end of a spear. Exceptions are morays that hang out at popular tourist dive sites and are relatively tame. In this case trust your local guide for advice, but always be cautious!

Species in the genus *Echidna* (94-96) have blunt teeth in contrast to most other eels. This is an adaptation for feeding on shelled molluscs and crustaceans. They exhibit striking colour patterns and are sometimes kept as aquarium pets.

The Painted Moray (107) sometimes frightens beachcombers. It occurs in very shallow pools at low tide or is occasionally found high and dry under rocks.

97
94
95
96
98
99
100
101
102
103
104
105
106
107
108

109 BLACK-EDGED CONGER
Conger cinereus Rüppell
Inhabits coral reef crevices; distinguished by well developed pectoral fins, relatively tall, black-edged dorsal and anal fins, and diagonal dark band behind mouth; Ningaloo Reef northwards; Indo-C. Pacific; to 103 cm.

110 MARBLED SNAKE-EEL
Callechelys marmoratus (Bleeker)
Inhabits sand bottoms near coral reefs; distinguished by dense pattern of irregular black spots; Ningaloo Reef northwards; Indo-C. Pacific; to 57 cm.

111 STARGAZER SNAKE-EEL
Brachysomophis cirrocheilos (Bleeker)
Inhabits sand bottoms, often with only eyes protruding above surface; distinguished by upward directed eyes near tip of snout, fringe of skin tentacles on lips, fang-like teeth in jaws and roof of mouth, and overall pale colour; Broome northwards; Indo-W. Pacific; to 125 cm.

112 SLENDER WORM-EEL
Muraenichthys gymnotus Bleeker
Inhabits sand bottoms near coral reefs; distinguished by small worm-like body, olive coloured back, pale belly and lack of pectoral fins; Ningaloo Reef northwards; Indo-C. Pacific; to 17 cm.

113 FRINGE-LIPPED SNAKE-EEL
Cirrhimuraena calamus (Günther)
Inhabits sand bottoms; distinguished by fringe of skin tentacles on upper lip, small pectoral fin, brownish colour of back, and abruptly pale on lower half; Geographe Bay northwards; W.A. only; to 62 cm.

114 CULVERIN
Leiuranus semicinctus (Lay & Bennett)
Inhabits sand bottoms; distinguished by series of black saddles on upper two-thirds of body; Ningaloo Reef northwards; Indo-C. Pacific; to 60 cm.

115 HARLEQUIN SNAKE-EEL
Myrichthys colubrinus (Boddaert)
Inhabits sand bottoms; similar to 114, but black bars completely or nearly encircle body; Abrolhos northwards; Indo-C. Pacific; to 88 cm.

116 FLAPPY SNAKE-EEL
Phyllophichthus xenodontus Gosline
Inhabits sand bottoms near reefs; distinguished by long pointed snout, leaf-like skin flap at each anterior nostril (near snout tip), and small pectoral fins; Albany northwards; Indo-C. Pacific; to 42 cm.

117 ONE-BANDED SNAKE-EEL
Ophichthus cephalozona Bleeker
Inhabits sand bottoms; distinguished by white-edged black saddle on middle of head; Onslow northwards; mainly W. Pacific; to 80 cm.

118 OLIVE SNAKE-EEL
Ophichthus rutiodermatoides (Bleeker)
Inhabits sand bottoms; distinguished by non-descript pattern, pointed snout, pointed teeth, and pectoral fin base on upper half of gill opening; Broome northwards; Indo-Australian Archipelago; to 68 cm.

119 BLACK-FINNED SNAKE-EEL
Ophichthus melanochir Bleeker
Inhabits sand bottoms near coral reefs; distinguished by black edge on dorsal fin, also pectoral fins sometimes entirely or partly black; Ningaloo Reef northwards; Indo-C. Pacific; to 80 cm.

120 ESTUARY SNAKE-EEL
Pisodonophis boro (Hamilton)
Inhabits sand or mud bottoms, often in estuaries or freshwater streams; distinguished by non-descript pattern, granular teeth, pectoral fin broad-based (not restricted to upper half of gill opening), and dorsal fin begins behind end of pectoral fins; Shark Bay northwards; Indo-W. Pacific; to 100 cm.

121 BURROWING SNAKE-EEL
Pisodonophis cancrivorous (Richardson)
Inhabits sand bottoms; distinguished by blunt snout (jaws equal in length), granular teeth, pectoral fin broad-based, and dorsal fin begins above pectoral fins; Fremantle northwards, but rare south of Shark Bay; Indo-C. Pacific; to 75 cm.

122 CHINGILT
Yirrkala lumbricoides (Bleeker)
Inhabits sand bottoms near coral reefs; distinguished by slender worm-like body, moderately long pointed snout, dorsal fin begins above gill openings, no pectoral fins, and anus about midway between snout and tip of tail; Ningaloo Reef northwards; Indo-W. Pacific; to 44 cm.

123 VULTURE EEL
Icththyapus vulturis Weber & de Beaufort
Inhabits sand bottoms near coral reefs; distinguished by general pale colouration, long pointed snout, very small eye and no pectoral fins; Ningaloo Reef northwards; Indo-C. Pacific; to 50 cm.

SNAKE EELS

All of the species on this plate, except no. 109, are members of the family Ophichthidae known as snake eels. Although they are very common most people, including keen anglers, are unaware of their presence. This is because they spend most of the time buried in the sand. Most of the species have a somewhat pointed snout to aid them in burrowing. In addition, many have a bony, sharp tail and are equally adept at burrowing forward or backward. The diet of most snake eels consists of small fishes, crabs, and prawns.

A few species, particularly 110, 114, and 115, are sometimes mistaken for sea snakes, but they can be easily distinguished by their lack of scales and possession of a pointed tail (paddle-like in snakes).

The Black-edged Conger Eel (109) belongs to the family Congidae. It is found in rocky areas and amongst coral reef crevices. In some parts of the Indo-Pacific region its flesh is considered a delicacy. Neither the conger or snake eels are dangerous.

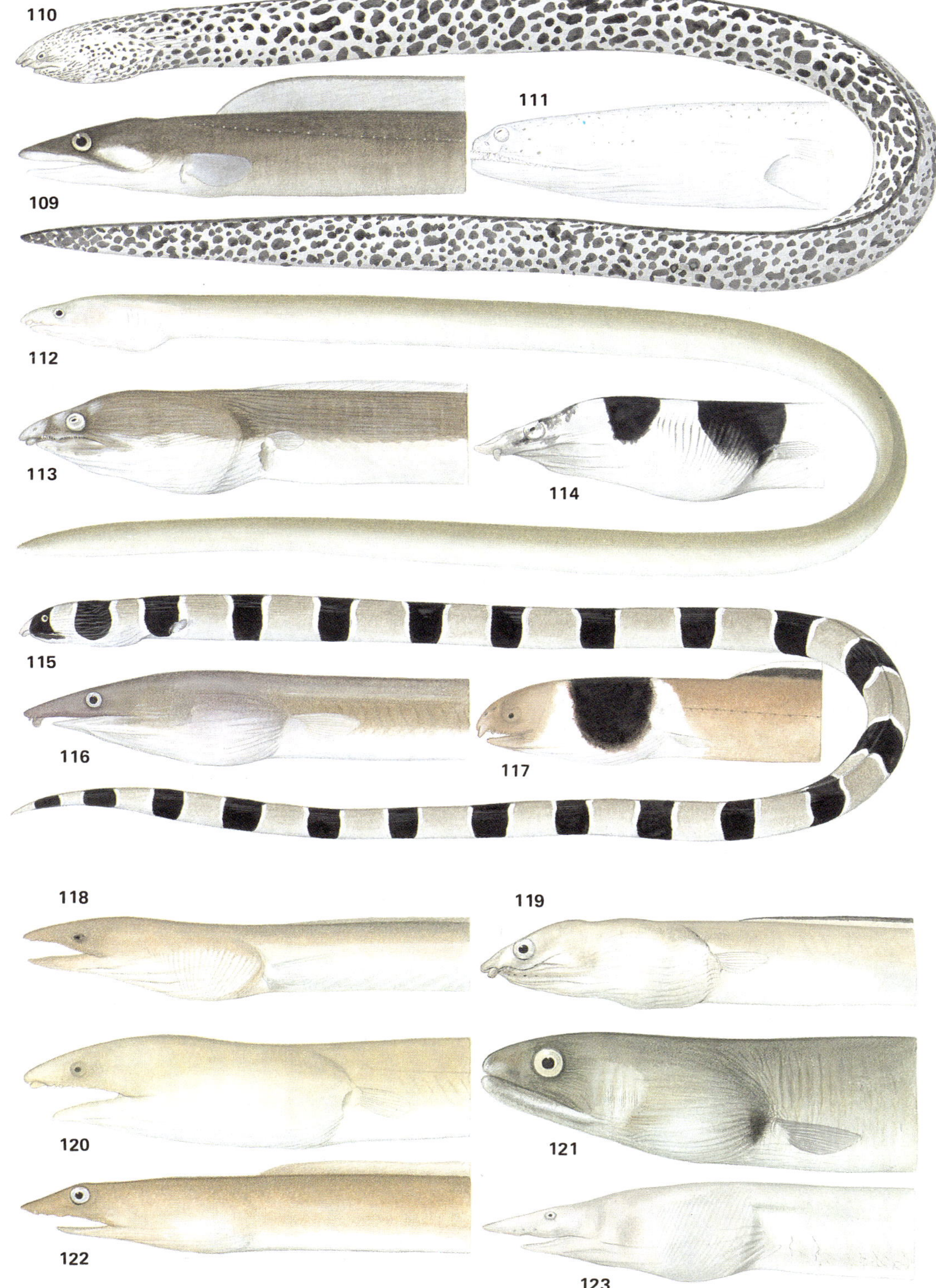

124 GIANT SALMON CATFISH
Arius thalassinus (Rüppell)
Inhabits coastal waters; there are about 8 species of salmon catfishes in northern waters which are difficult to identify even for experts; *A. thalassinus* is the largest species and has 6 patches of teeth on roof (palate) of mouth (versus 2-4 in others); Abrolhos northwards; Indo-C. Pacific; to 185 cm; 4.55 kg. ★★★

125 SMALLER SALMON CATFISH
Arius graefei Kner & Steindachner
Inhabits coastal waters, estuaries, and freshwater streams; difficult to identify, but has gill raker-like processes on back of all gill arches, 4 oval patches of teeth on roof of mouth, and barbel on upper jaw not reaching farther than beginning of dorsal fin; Exmouth Gulf northwards; N. Australia and New Guinea; to 50 cm. ★★★

126 NAKED-HEADED CATFISH
Euristhmus nudiceps (Günther)
Inhabits coastal waters; similar to 127, but is more slender with greatest depth of body (i.e. height of body below first dorsal fin) fitting about 10 to 13 times in total length; Shark Bay northwards; N. Australia and New Guinea; to 33 cm.

127 LONG-TAILED CATFISH
Euristhmus lepturus (Günther)
Inhabits coastal waters; similar to 126 but not quite as slender with greatest depth of body (i.e. height of body below first dorsal fin) fitting about 8 or 9 times in total length; Exmouth Gulf northwards; N. Australia and New Guinea; to 46 cm.

128 STRIPED CATFISH
Plotosus lineatus (Thunberg)
Inhabits coastal waters, frequently in the vicinity of coral reefs; juveniles may form tightly packed aggregations containing up to several hundred fish; nearly entire coast of W.A.; Indo-C. Pacific; to 32 cm. ★★★

129 WHITE-LIPPED CATFISH
Paraplotosus albilabris (Valenciennes)
Inhabits coastal reefs, frequently found amongst weed; distinguished from 130, by lighter colour and much shorter dorsal fin; Cape Leeuwin northwards; Indo-Australian Archipelago; to 134 cm; .95 kg. ★★★

130 SAILFIN CATFISH
Paraplotosus sp.
Inhabits coastal reefs, usually in the vicinity of coral reefs; distinguished by black colour and tall dorsal fin; Point Quobba northwards; W.A. only; to 30 cm.

131 INDIAN LIZARDFISH
Synodus indicus (Day)
Inhabits trawling grounds; distinguished by overall light colour with faint stripes on back and 2 dark streaks on upper corner of gill cover; Broome northwards; Indian Ocean and Indo-Australian Archipelago; to 21 cm.

132 TAILSPOT LIZARDFISH
Synodus jaculum Russell & Cressey
Inhabits the vicinity of coral reefs; distinguished by black spot at base of tail; Ningaloo Reef northwards; Indo-C. Pacific; to 13 cm.

133 BLACK LIZARDFISH
Synodus kaianus (Günther)
Inhabits trawling grounds; distinguished by overall dark colouration; Derby northwards; mainly W. Pacific; to 22 cm.

134 BIG-EYED LIZARDFISH
Synodus macrops Tanaka
Inhabits trawling grounds; distinguished by large size of eye and 3 large dark blotches on side; Broome northwards; Andaman Sea and W. Pacific; to 20 cm.

135 BLACK-SHOULDERED LIZARDFISH
Synodus hoshinonis Tanaka
Inhabits trawling grounds; distinguished by prominent black area on upper edge of gill cover; Onslow northwards; Indo-W. Pacific; to 22 cm.

136 NETTED LIZARDFISH
Synodus sageneus Waite
Inhabits trawling grounds; distinguished from other lizardfishes by the absence or reduced size of the adipose fin (small fin on back between dorsal fin and tail); also known as Fishnet lizardfish; Fremantle northwards; E. Indian Ocean and Indo-Australian Archipelago; to 26 cm; .15 kg.

137 VARIEGATED LIZARDFISH
Synodus variegatus (Lacepède)
Inhabits sand rubble areas in the vicinity of coral reefs; distinguished by mottled appearance with series of dark bars on side; Jurien Bay northwards; Indo-C. Pacific; to 25 cm; .24 kg.

138 PAINTED GRINNER
Trachinocephalus myops (Schneider)
Inhabits coastal waters and trawling grounds; distinguished by pug-headed appearance, yellowish colour and bluish stripes on side; Fremantle northwards; Indo-C. Pacific; to 66 cm; 1.588 kg. ★★

139 SLENDER GRINNER
Saurida gracilis (Quoy & Gaimard)
Inhabits sandy areas, frequently near coral reefs; similar to 137, but teeth not covered by lips when mouth is closed; Ningaloo Reef northwards; Indo-C. Pacific; to 28 cm.

140 LARGE-SCALED GRINNER
Saurida undosquamis (Richardson)
Inhabits trawling grounds; distinguished by small black spots along upper edge of tail; also known as Checkered lizardfish; nearly entire coast of W.A.; Indo-W. Pacific; to 45 cm; 1.15 kg. ★★

141 COMMON GRINNER
Saurida tumbil (Bloch)
Inhabits trawling grounds; distinguished by lack of markings and dark lower lobe of tail; Shark Bay northwards; Indo-W. Pacific; to 43 cm. ★★

142 GLASSY BOMBAY DUCK
Harpodon translucens Saville-Kent
Inhabits bays and estuaries; distinguished by large curved teeth and flaccid semi-transparent appearance; Derby northwards; N. Australia only; to 70 cm. ★★

36

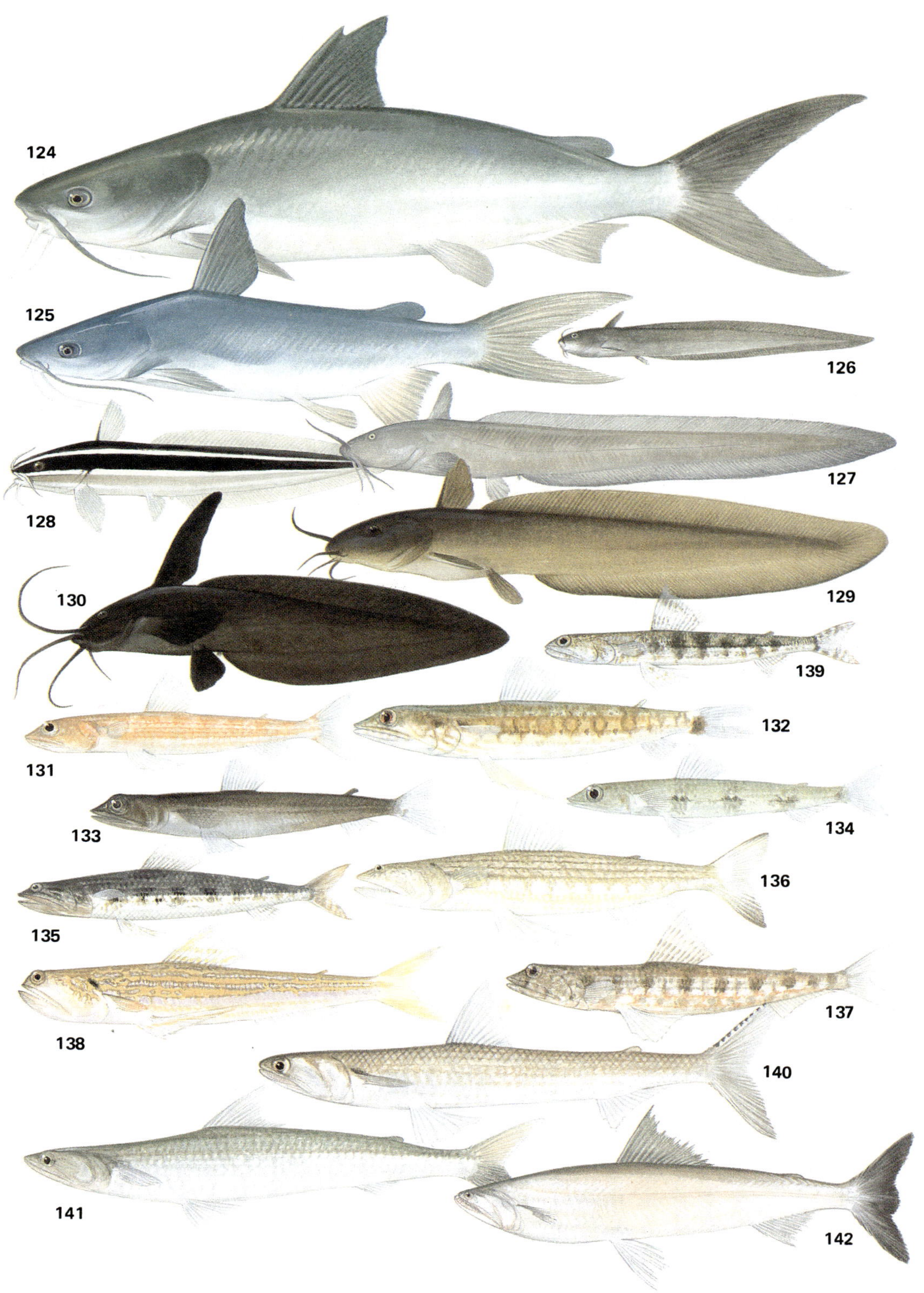

143 DAHL'S FROGFISH
Batrachomoeus dahli (Rendahl)
Inhabits shallow reefs and offshore trawling grounds; distinguished by gill slit extending along entire pectoral fin base and marbled pattern without distinctive cross bars; Shark Bay to Broome; W.A. only; to 20 cm.

144 WESTERN FROGFISH
Batrachomoeus occidentalis Hutchins
Inhabits offshore trawling grounds; distinguished by gill slit extending along entire pectoral base and distinct cross bars on side; Rottnest Island to Exmouth Gulf; W.A. only; to 20 cm.

145 THREE-SPINED FROGFISH
Batrachomoeus trispinosus (Günther)
Inhabits reefs, mangrove estuaries, and offshore trawling grounds; similar to 143, but has relatively distinct cross bars on side and well contrasted markings on dorsal surface of head; Barrow Island northwards; N. Australia and New Guinea; to 30 cm.

146 BANDED FROGFISH
Halophyrne diemensis (Leseuer)
Inhabits reef crevices and offshore trawling grounds; distinguished by gill slit extending only half to two-thirds of pectoral fin base and lacks small dark-edged white spots on side; Shark Bay northwards; Indo-Australian Archipelago; to 26 cm; 0.4 kg.

147 OCELLATED FROGFISH
Halophyrne ocellatus Hutchins
Inhabits offshore trawling grounds, may enter craypots; distinguished by restricted gill slit as in 146, but has small dark-edged white spots on side; Fremantle to Broome; W.A. only; to 26 cm.

148 URCHIN CLINGFISH
Diademichthys lineatus (Sauvage)
Inhabits coral reefs amongst spines of sea urchins or in branching corals; distinguished by peculiar shape with pale stripe on middle of side; Ningaloo Reef northwards; Indo-W. Pacific; to 5 cm.

149 SHARK BAY CLINGFISH
Lepadichthys sandracatus Whitley
Inhabits trawling grounds; distinguished by red to brown colour, dark stripe behind eye, and peculiar shape with suction-disk on belly; Port Gregory to Shark Bay; W.A. only; to 5 cm.

150 HUMPBACK ANGLERFISH
Tetrabrachium ocellatum (Günther)
Inhabits offshore trawling grounds; distinguished by amorphous appearance, low dorsal and anal fins, small eyes and mouth, and numerous white spots; Dongara northwards; N. Australia and New Guinea; to 8.5 cm.

151 STRIPED ANGLERFISH
Antennarius striatus Shaw
Inhabits inshore reefs, often in weeds; distinguished by narrow dark streaks and elongate blotches on body and fins; also has entirely black variety; anglerfishes lure small fishes by wriggling an enticing "bait" on the tip of the first dorsal spine; Abrolhos and Geraldton northwards; Indo-W. Pacific; to 22 cm; .15 kg.

152 WHITE-FINGER ANGLERFISH
Antennarius nummifer (Cuvier)
Inhabits inshore reefs; similar to 153, but has distinct tail base and usually with dark spot at base of dorsal fin; Jurien Bay northwards; Indo-W. Pacific; to 13 cm.

153 FRECKLED ANGLERFISH
Antennarius coccineus (Lesson)
Inhabits offshore coral reefs; similar to 152, but lacks distinct tail base and black spot on dorsal fin base; North West Shelf; Indo-E. Pacific; to 13 cm.

154 SHAGGY ANGLERFISH
Antennarius hispidus (Bloch & Schneider)
Inhabits coastal reefs and offshore trawling grounds; distinguished by scattered dark blotches on side and fins, and diagonal, elongate streaks and blotches on dorsal fin; Broome northwards; Indo-W. Pacific; to 20 cm.

155 PAINTED ANGLERFISH
Antennarius pictus (Shaw & Nodder)
Inhabits coastal reefs; distinguished by scattered dark spots (some pale-edged) and with light patches on nape, cheek, and pectoral regions; Geraldton northwards; Indo-W. Pacific; to 24 cm.

156 SPOTTED-TAIL ANGLERFISH
Antennarius trisignatus (Richardson)
Inhabits coastal reefs, also found under wharves; distinguished by dark-edged pale spots on tail; Shark Bay northwards; Indo-Australian Archipelago; to 18 cm.

157 SARGASSUM FISH
Histrio histrio (Linnaeus)
Inhabits oceanic waters and inshore reefs, usually found in clumps of floating sargassum weed; distinguished by smooth skin (most other anglerfishes have prickles), often with skin flaps and filaments on head and body, and limb-like pectoral fins; also known as Marbled angler; Rockingham northwards; worldwide temperate and tropical seas; to 16 cm.

158 RED CUSKEEL
Ogilbia sp.
Inhabits coral reef crevices; distinguished by yellow to orange colour, elongate, tapering shape, blunt snout, and tail separated from dorsal and anal fins; Ningaloo northwards; possibly Australia only; to 9 cm.

159 PEARLFISH
Onuxodon margaritiferae (Rendahl)
Inhabits coastal waters, lives in the mantle cavity of oysters; other similar species found inside sea cucumbers and cushion starfish; distinguished by long slender, transparent body; Barrow Island northwards; W. Australia only; to 9 cm.

160 BLACK-EDGED CUSKEEL
Ophidion muraenolepis (Günther)
Inhabits trawling grounds; distinguished by long tapering body, low, dark edged dorsal and anal fins, no barbels, but thread-like pelvic fins just behind chin; Onslow northwards; Indo-Australian Archipelago; to 20 cm.

161 BEARDED CUSKEEL
Brotula multibarbata Temminck & Schlegel
Inhabits coastal reefs and deeper offshore waters; distinguished by robust, elongate and tapering body, barbels around mouth, and thread-like pelvic fins below edge of gill cover; Barrow Island northwards; Indo-C. Pacific; to 90 cm. ★★

162 GOLDEN CUSKEEL
Sirembo imberis (Temminck & Schlegel)
Inhabits trawling grounds; similar to 160, but with broken stripes and spots on side; Onslow northwards; N. Indian Ocean and W. edge of Pacific; to 20 cm.

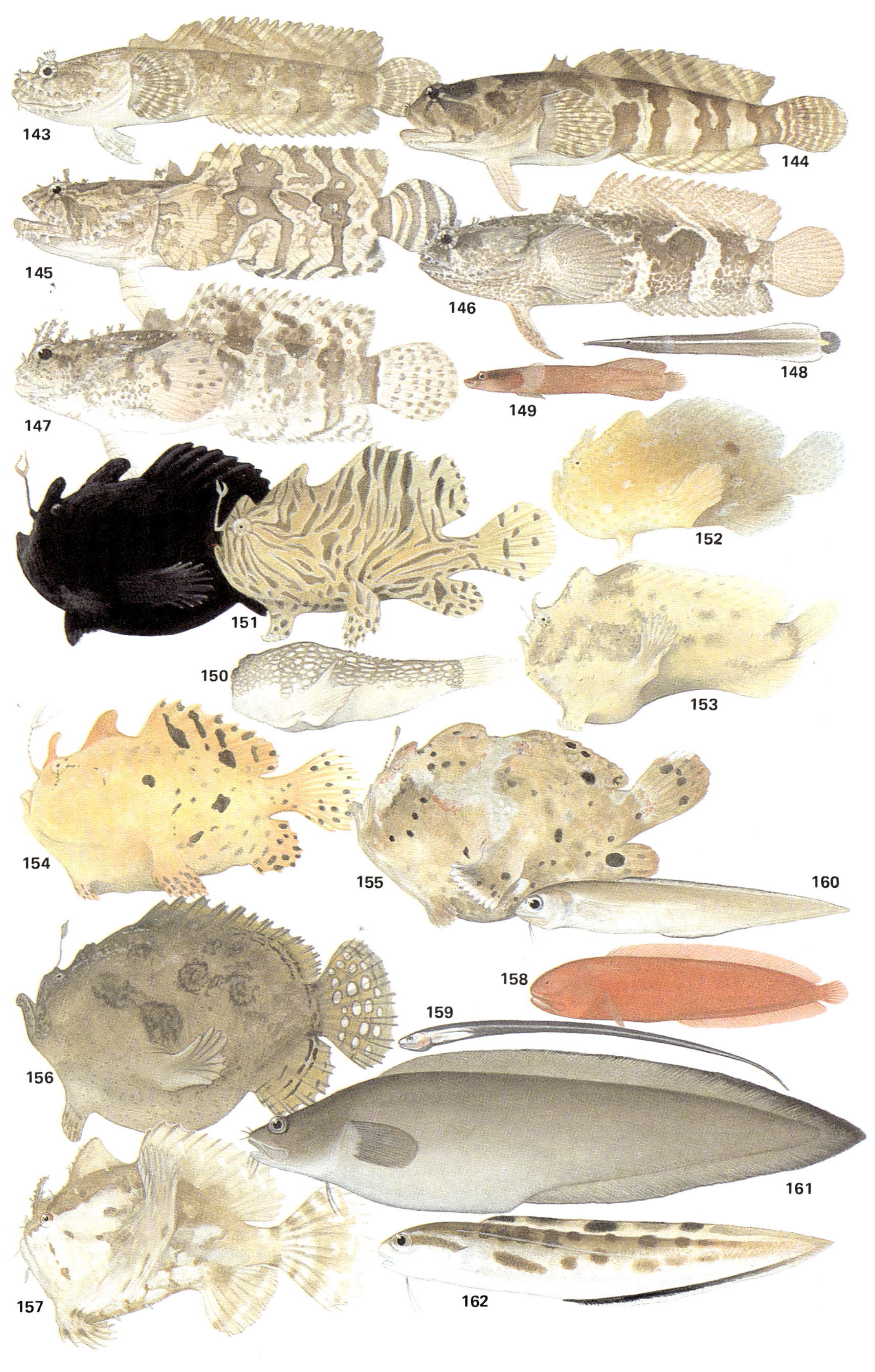

163 FLYINGFISH
Cypselurus sp.
Inhabits oceanic waters, frequently well offshore; several species of flyingfishes are common in north-western waters, but only one is shown here as they are seldom caught by anglers and identification at the species level is frequently difficult; flyingfishes are distinguished by the wing-like pectoral fins and elongated lower tail lobe which facilitate long gliding flights over the sea surface; entire coast of W.A.; to 27 cm. ★★

164 SNUB-NOSED GARFISH
Arramphus sclerolepis Günther
Inhabits coastal waters, sometimes entering brackish estuaries and the lower reaches of freshwater streams; distinguished by its short snout and lack of elongated lower jaw (as in other garfishes); Carnarvon northwards; N. Australia and New Guinea; to 38 cm. ★★★

165 BUFFON'S GARFISH
Zenarchopterus buffonis (Valenciennes)
Inhabits coastal waters, sometimes entering estuaries; distinguished by truncate tail (forked in other garfishes) and prominent dark brown stripe along midline of snout; male (shown here) has elongate and thickened anal fin; Port Hedland northwards; Andaman Sea and Indo-Australian Archipelago; to 13 cm. ★★

166 BARRED GARFISH
Hemirhamphus far (Forsskål)
Inhabits coastal waters, frequently in schools near reefs; distinguished by 4-6 prominent dark bars on side; Dampier Archipelago northwards; Indo-W. Pacific; to 35 cm. ★★★

167 ROBUST GARFISH
Hemirhamphus robustus Günther
Inhabits coastal waters; similar to 166, but lacks dark bars on side (may have a single dark blotch below dorsal fin), both these species differ from 168 and 169 in lacking scales on the triangular-shaped surface of the upper jaw; also known as Three by Two garfish; Geographe Bay northwards; Australia only; to 48 cm. ★★★

168 TROPICAL GARFISH
Hyporhamphus affinis (Günther)
Inhabits coastal waters, occurring in schools; similar to 166 and 167, but has scales (versus no scales) on the triangular-shaped surface of the upper jaw, is more slender, and lacks bars or blotches on side; Shark Bay northwards; Indo-C. Pacific; to 25 cm. ★★

169 QUOY'S GARFISH
Hyporhamphus quoyi (Valenciennes)
Inhabits coastal waters, occurring in schools, distinguished by the relatively short lower jaw; Shark Bay northwards; E. Indian Ocean and W. Pacific; to 34 cm. ★★★

170 LONG-FINNED GARFISH
Euleptorhamphus viridis (van Hasselt)
Inhabits coastal waters, but sometimes encountered well offshore; distinguished by strongly compressed ribbon-like body, very long lower jaw, and wing-like pectoral fins; exhibits gliding behaviour similar to flyingfishes; Albany northwards; Indo-E. Pacific; to 60 cm. ★★★

171 BARRED LONGTOM
Ablennes hians (Valenciennes)
Inhabits oceanic waters, often well offshore; distinguished by bars on rear part of body below dorsal fin; Fremantle northwards; worldwide tropical and subtropical seas; to 120 cm. ★★

172 FLAT-TAILED LONGTOM
Platybelone platyura (Bennett)
Inhabits offshore waters; distinguished by flattened (dorsoventrally) tail base; Point Quobba northwards; Indo-C. Pacific; to 40 cm. ★★

173 SLENDER LONGTOM
Strongylura leiura (Bleeker)
Inhabits coastal waters, frequently in bays and estuaries; distinguished by elongate jaws, slender shape, and black bar or streak across base of gill cover (not shown); Albany northwards; Indo-W. Pacific; to 110 cm. ★★

174 CROCODILIAN LONGTOM
Tylosurus crocodilus (Peron & Le Sueur)
Inhabits coastal waters; distinguished by dark fleshy ridge on side of tail base (not shown); Ningaloo Reef northwards; Atlantic and Indo-C. Pacific; to 150 cm. ★★

175 STOUT LONGTOM
Tylosurus gavialoides (Castelnau)
Inhabits coastal waters; similar to 174, but lacks fleshy ridge on side of tail base, also has more rounded snout tip (when viewed from above); Shark Bay northwards; E. Indian Ocean and W. Pacific; to 130 cm. ★★

176 SPOTTED HARDYHEAD
Allanetta mugiloides (McCulloch)
Inhabits coastal waters, frequently in schools off beaches; distinguished by small dark spot below pectoral fin; Dirk Hartog Island northwards; N. Australia only; to 6 cm.

177 FEW-RAYED HARDYHEAD
Craterocephalus pauciradiatus (Günther)
Inhabits coastal waters, occurring in schools; a non-descript silvery fish which has the anus positioned close to the pelvic fin base, and lacks a notch on the lower cheek margin; Abrolhos northwards; N. Australia only; to 6 cm.

178 SAMOAN HARDYHEAD
Hypoatherina temminckii (Bleeker)
Inhabits coastal waters; distinguished by slender shape, anus placed far behind pelvic fin base, and prominent silvery midlateral stripe; Ningaloo Reef northwards; Indo-C. Pacific; to 10 cm.

179 ENDRACHT HARDYHEAD
Pranesus endrachtensis (Quoy & Gaimard)
Inhabits coastal waters; similar to 176, but no spot below pectoral fin and rear part of lower jaw only slightly elevated (versus prominently elevated —note: mouth must be opened widely to detect this feature); Lancelin northwards; N. Australia and New Guinea; to 10 cm.

180 OGILBY'S HARDYHEAD
Pranesus ogilbyi Whitley
Inhabits shallow coastal waters, including bays and estuaries, usually in schools; distinguished by dark blotch at tip of pectoral fins; Geographe Bay northwards; Australia only; to 17 cm.

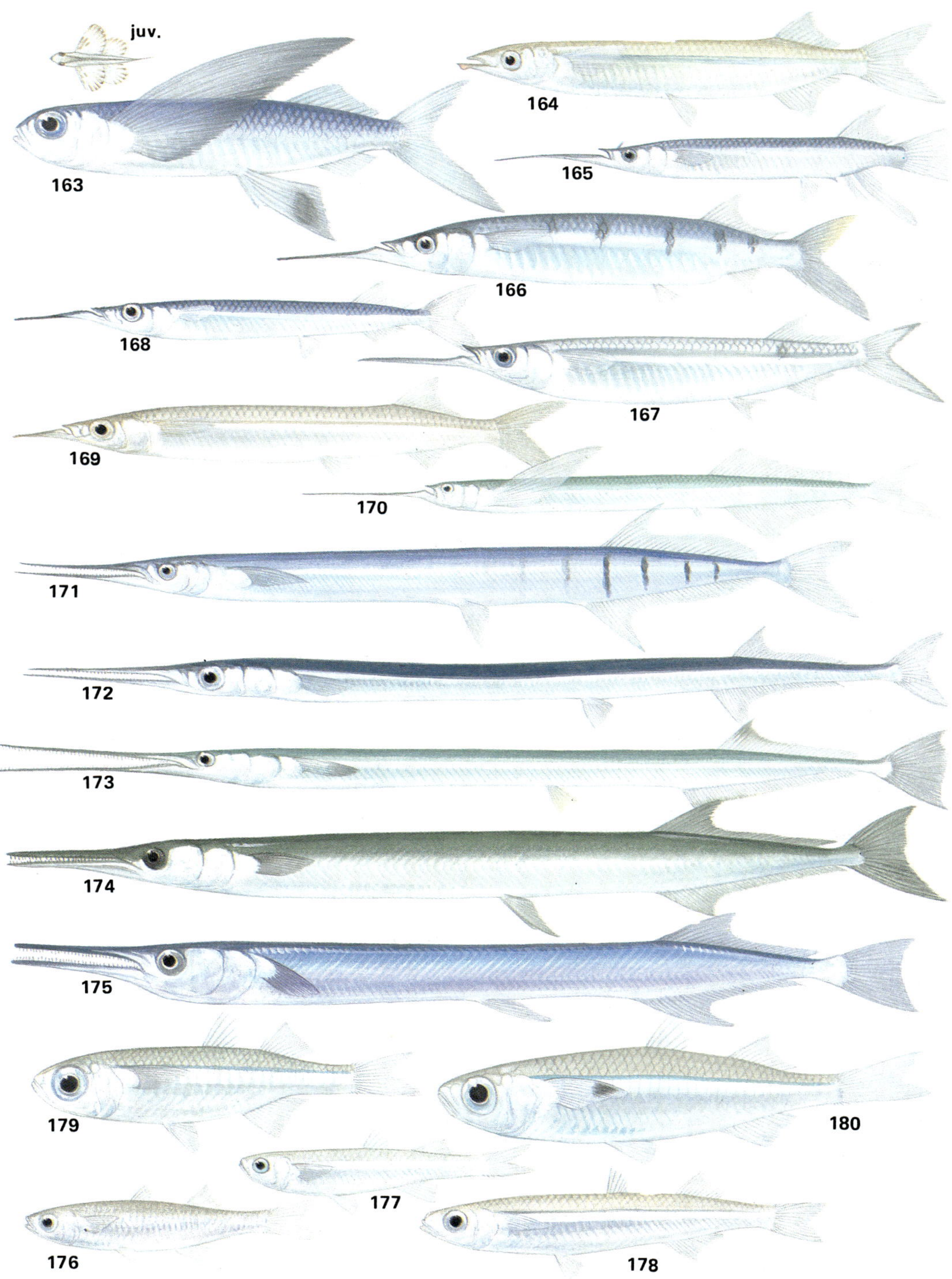

181 BLACKFIN SQUIRRELFISH
Myripristis adustus Bleeker
Inhabits caves and crevices of coral reefs; distinguished by black area on outer half of dorsal and anal fins; North West Shelf; Indo-W. Pacific; to 30 cm. ★★★

182 DOUBLETOOTH SQUIRRELFISH
Myripristis hexagonatus (Lacepède)
Inhabits caves and crevices of coral reefs; similar to 184 (both have 2 pairs of tooth patches on front of lower jaw, just outside of the mouth), but differs in having small scales on the inside base ("armpit") of pectoral fin; Dampier Archipelago northwards; Indo-W. Pacific; to 20 cm. ★★★

183 KUNTEE SQUIRRELFISH
Myripristis kuntee Cuvier
Inhabits caves and crevices of coral reefs; distinguished by broad dark behind head to pectoral fin base and 37-44 scales in lateral line; Ningaloo Reef northwards; Indo-C. Pacific; to 20 cm. ★★★

184 PALE SQUIRRELFISH
Myripristis melanostictus Bleeker
Inhabits caves and crevices of coral reefs; similar to 182, but differs in lacking small scales on the inside base ("armpit") of pectoral fin; Dampier Archipelago northwards; Indo-W. Pacific; to 30 cm. ★★★

185 CRIMSON SQUIRRELFISH
Myripristis murdjan (Forsskål)
Inhabits caves and crevices of coral reefs; distinguished by dark margin on upper part of gill cover and 27-32 scales in lateral-line; Point Quobba northwards; Indo-C. Pacific; to 30 cm; .291 kg. ★★★

186 SPOTFIN SQUIRRELFISH
Neoniphon sammara (Forsskål)
Inhabits patch reefs in lagoons amongst branching corals; distinguished by slender shape, silvery or pale colouration, and black spot at front of dorsal fin; Ningaloo Reef northwards (rare); Indo-C. Pacific; to 32 cm. ★★

187 DEEPWATER SQUIRRELFISH
Ostichthys kaianus (Günther)
Inhabits offshore trawling ground between about 300-650 m depth; distinguished by broad V-shaped groove at middle of snout and 12 dorsal spines; North West Shelf; Indo-W. Pacific; to 30 cm. ★★

188 ROUGH SQUIRRELFISH
Pristilepis oligolepis (Whitley)
Inhabits offshore trawling grounds; distinguished by narrow, elongate groove at middle of snout and 12 dorsal spines; Shark Bay northwards; mainly W. Pacific; to 20 cm. ★★

189 CROWNED SQUIRRELFISH
Sargocentron diadema (Lacepède)
Inhabits coral reefs to 30 m depth; distinguished by broad black margin on dorsal fin; North West Shelf; Indo-C. Pacific; to 16 cm. ★★

190 SPECKLED SQUIRRELFISH
Sargocentron punctatissimum (Cuvier)
Inhabits rocky shores and reefs exposed to wave action; distinguished by pepper-like spotting on sides; North West Shelf; Indo-C. Pacific; to 15 cm. ★★

191 RED SQUIRRELFISH
Sargocentron rubrum (Forsskål)
Inhabits live and dead coral reefs, usually in protected lagoons; distinguished by red and white stripes of about equal width; Point Quobba northwards; Indo-W. Pacific; to 28 cm. ★★

192 SPINY SQUIRRELFISH
Sargocentron spiniferum (Forsskål)
Inhabits caves and crevices of coral reefs; distinguished by overall red colouration and very long spine at lower margin of cheek; North West Shelf; Indo-C. Pacific; to 40 cm. ★★★

193 BLUESTRIPE SQUIRRELFISH
Sargocentron tiere (Cuvier)
Inhabits mainly outer exposed reefs to 20 m depth; distinguished by brilliant red colour, and iridescent blue stripes on lower sides; Ningaloo Reef northwards; Indo-C. Pacific; to 34 cm. ★★★

194 VIOLET SQUIRRELFISH
Sargocentron violaceus (Bleeker)
Inhabits caves and crevices of coral reefs; distinguished by overall dusky appearance; Dampier Archipelago northwards; Indo-W. Pacific; to 23 cm. ★★

PREDATORS OF THE NIGHT

The fishes featured on this plate are members of the family Holocentridae, commonly known as squirrelfishes or soldierfishes. They occur in all tropical seas. Most of the approximately 70 species inhabit the Indo-Pacific region. They are characterised by rough scales, prominent fin spines, a large eye, and red coloration. They are primarily nocturnal, sheltering in caves and under ledges during the day. They are one of the most common groups of fishes seen by divers who visit the reef at night. They emerge from their retreats at dusk and spend the night feeding out in the open on such items as fishes and crustaceans. Many species, particularly 189-194, possess a sharp spine at the back of each cheek, which can inflict a painful wound if handled carelessly. Therefore caution must be exercised when removing them from a hook. Although most of the species are small, the flesh is considered good eating.

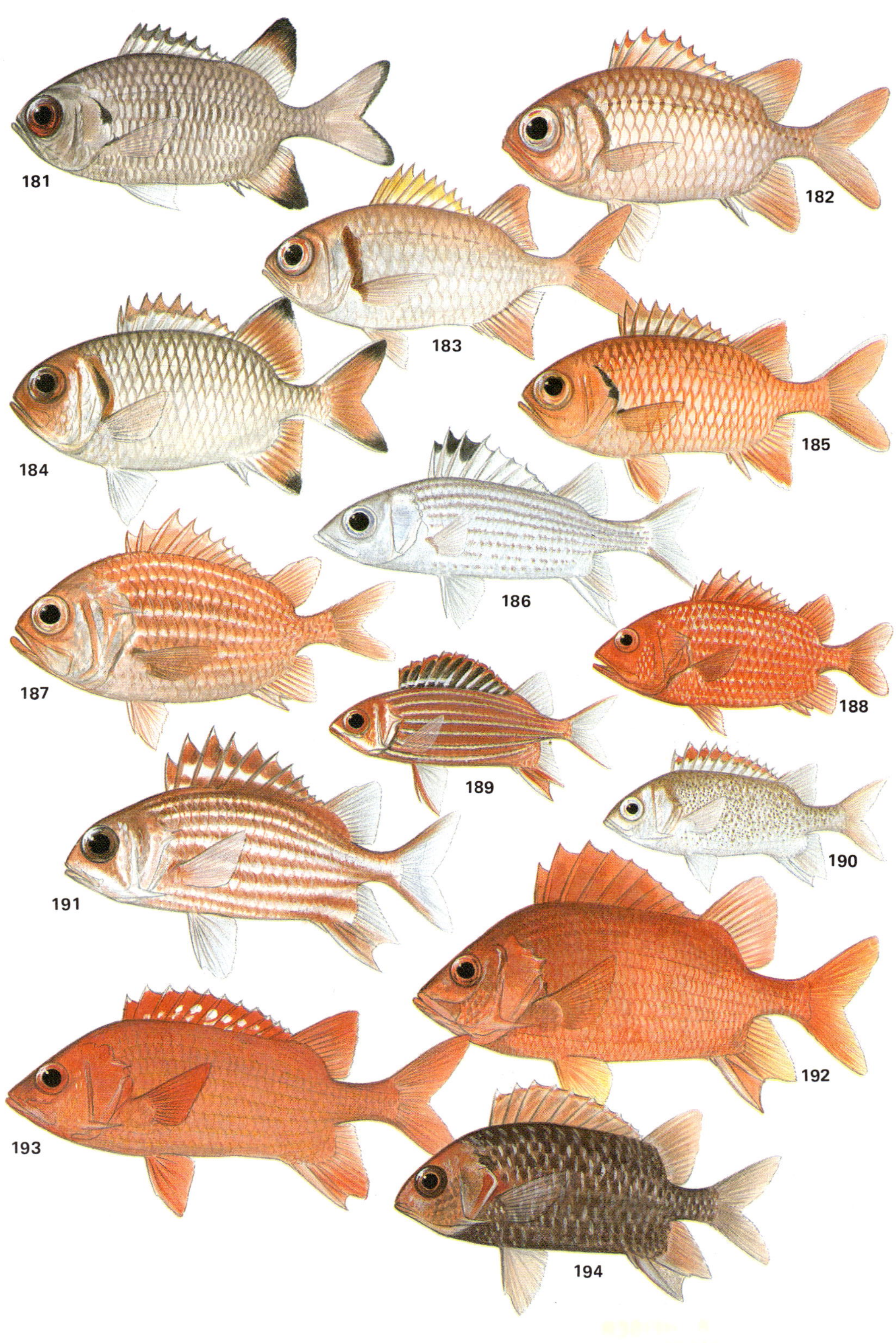

195 KNIGHT FISH
Cleidopus gloriamaris De Vis
Inhabits coastal reefs, usually in caves or under ledges; similar to 196, but has very conspicuous light-producing organ on each side of lower jaw (appearing as orange spot in daylight or a blue-green one at night), and scales are more strongly outlined by dark colouration; also known as Pineapple fish; mainly southern waters ranging north to Shark Bay; Australia only; to 28 cm.

196 JAPANESE PINEAPPLEFISH
Monocentris japonica (Houttuyn)
Inhabits deeper offshore reefs and trawling grounds; similar to 195, but light organs not as conspicuous, narrower dark margins around scales, and wider gap between eye and mouth; Onslow northwards; N. Indian Ocean and W. Pacific; to 20 cm.

197 LITTLE DORY
Cyttopsis cypho (Fowler)
Inhabits deeper trawling grounds of the continental shelf; general shape similar to 198 and 199, but a much smaller fish lacking filamentous dorsal fin rays; Broome northwards; mainly W. Pacific; to 18 cm.

198 MIRROR DORY
Zenopsis nebulosus (Temminck & Schlegel)
Inhabits deeper trawling grounds of the continental shelf; similar to 199, but lacks scales (versus small scales present) and forehead profile distinctly concave (versus convex); entire coast of W.A.; C. and W. Pacific; to 58 cm. ★★★★

199 JOHN DORY
Zeus faber Linnaeus
Inhabits deeper trawling grounds of the continental shelf, although sometimes found close to the coast; similar to 198, but has small scales (versus no scales) and forehead profile is concave (versus convex); entire coast of W.A.; tropical and temperate E. Atlantic and Indo-W. Pacific; to 75 cm. ★★★★

200 PINK BOARFISH
Antigonia rhomboidea McCulloch
Inhabits deeper trawling grounds of the continental shelf; distinguished by diamond-shaped body; red or pink colouration with yellowish fins; Abrolhos northwards; Australia only; to 15 cm.

201 HIGH-FINNED VEILFIN
Velifer hypselopterus Bleeker
Inhabits deeper trawling grounds of the continental shelf, although the young may appear in shallow coastal waters; distinguished by filamentous dorsal and anal fins and diffuse dark bars on side; *V. multiradiatus* (not shown) is similar, but has more dorsal and anal fin rays (41-44 versus 32-36 dorsal rays and 33-36 versus 25-27 anal rays); Shark Bay northwards; Indo-W. Pacific; to 40 cm.

202 SMOOTH FLUTEMOUTH
Fistularia commersonii Rüppell
Inhabits coastal waters in the vicinity of reefs; distinguished by long snout, trailing filament on tail, and greenish-brown colour of back; Cockburn Sound northwards; Indo-E. Pacific; to 163 cm.

203 ROUGH FLUTEMOUTH
Fistularia petimba Lacepède
Inhabits coastal waters, also found well offshore; similar to 202, but has row of bony plates along middle of back (absent in 202) and is reddish or brownish-orange in colour (versus greenish-brown); entire coast of W.A.; Atlantic and Indo-C. Pacific; to 185 cm.

204 PAINTED FLUTEMOUTH
Aulostomus chinensis (Linnaeus)
Inhabits coral reefs; roughly similar shape to 202 and 203, but has shorter snout, has small scales (versus no scales), row of feeble dorsal spines on back, different shaped fins, and lacks tail filament; colour is generally brown or greenish with faint pale stripes on side and white spots on rear part of body; in addition a yellow variety is frequently seen; Fremantle northwards; Indo-E. Pacific; to 50 cm.

205 GHOST FLATHEAD
Hoplichthys regani Jordan & Richardson
Inhabits deep offshore trawling grounds; similar to flatheads (273-281) in body shape (i.e. greatly flattened), but lacks scales and has filamentous dorsal-fin rays; Broome northwards; Indo-W. Pacific; to 20 cm.

206 ORIENTAL SEAROBIN
Dactyloptaenia orientalis (Cuvier)
Inhabits coastal waters, usually on sand bottoms near coral reefs; distinguished by huge wing-like pectoral fins; other similar species also occur off N.W. Australia and are best distinguished by their "wing" patterns: *D. macracanthus* — black blotch encompassing several yellow spots in centre of "wing", *D. papilio* — large black patch containing blue spots at base of "wing", and *D. peterseni* — "wing" dark green at centre crossed by many broken pale yellow lines and scattered spots; Geraldton northwards; Indo-C. Pacific; to 38 cm.

207 SLENDER SEAMOTH
Pegasus volitans Linnaeus
Inhabits sand or silt bottoms of bays and estuaries; distinguished by flattened head and tapered body encased in plate-like armour similar to seahorses, and fan-like pectoral fins; often identified as *Parapegasus natans;* Fremantle northwards; Indo-W. Pacific; to 16 cm.

208 SHORT SEAMOTH
Eurypegasus draconis (Linnaeus)
Inhabits sand or silt bottoms, frequently in bays or estuaries; similar to 207, but wider body (when viewed from above), shorter snout and tail, and body is more "sculptured"; Port Hedland northwards; Indo-W. Pacific; to 10 cm.

PLATE 14

209 GROOVED RAZORFISH
Centriscus scutatus Linnaeus
Inhabits coastal waters; distinguished by long snout and thin (highly compressed) body composed of bony plates; reported off Kimberley coast, but rare; Indo-W. Pacific; to 15 cm.

210 HARLEQUIN GHOST PIPEFISH
Solenostomus armatus Weber
Inhabits inshore reefs and weed beds, sometimes in floating seaweed; distinguished by skin flaps on head and body, stripes on body and spotted fins; Point Quobba northwards; Indo-W. Pacific; to 12 cm.

211 GHOST PIPEFISH
Solenostomus cyanopterus Bleeker
Inhabits inshore reef areas; distinguished by yellowish to green or brown colour, general shape similar to 210, but has fewer skin flaps and shorter tail base; Ningaloo Reef northwards; Indo-W. Pacific; to 16 cm.

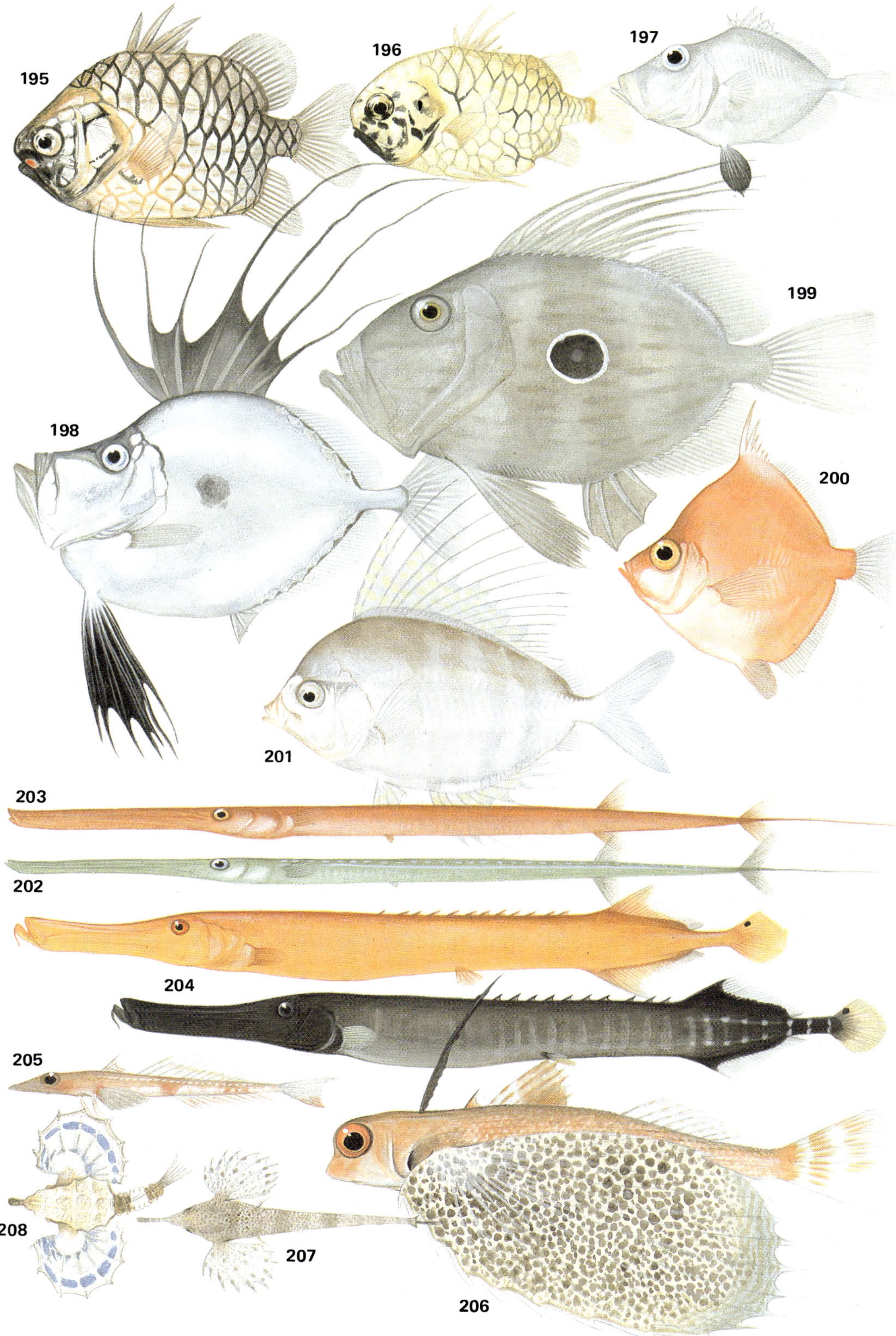

209 GROOVED RAZORFISH
Centriscus scutatus Linnaeus. Text on page 44.

210 HARLEQUIN GHOST PIPEFISH
Solenostomus armatus Weber. Text on page 44.

211 GHOST PIPEFISH
Solenostomus cyanopterus Bleeker. Text on page 44.

212 SPOTTED SEAHORSE
Hippocampus kuda Bleeker
Inhabits sheltered bays and estuaries; variable in colour, either yellow, brown or blackish, sometimes with dark bands or either dark or white spots, lacks narrow lines across snout found on 213; Onslow northwards; Indo-C. Pacific; to 30 cm.

213 WESTERN AUSTRALIAN SEAHORSE
Hippocampus angustatus Günther
Inhabits sheltered bays; variable in colour, either white, orange or yellow with light patches and sometimes network of fine dark lines, also has narrow lines across snout; Augusta to North West Cape; W.A. only; to 22 cm (illustration not shown at proper scale).

214 SPINY SEAHORSE
Hippocampus hystrix Kaup
Inhabits coastal waters; distinguished by pronounced spiny ridges on head and body; Onslow northwards; Indo-W. Pacific; to 15 cm.

215 LARSON'S PIPEHORSE
Acentronurus larsonae Dawson
Inhabits coral reefs or amongst sargassum weed; distinguished by seahorse shape and bulbous forehead; known only from Monte Bello Islands; to 3.5 cm.

216 EEL PIPEFISH
Bulbonaricus brauni (Dawson & Allen)
Inhabits coral reefs amongst organ-pipe coral (*Galaxea musicalis*); distinguished by eel-like shape and lack of fins; Ningaloo Reef; mainly Indo-Australian Archipelago; to 6 cm.

217 MURION PIPEFISH
Choeroichthys latispinosus Dawson
Inhabits coral reefs; distinguished by relatively broad, elongate snout with upturned mouth, and short tail; known thus far only from South Murion Island; to 3 cm.

218 SHORT-BODIED PIPEFISH
Choeroichthys brachysoma (Bleeker)
Inhabits reefs and seagrass beds; distinguished by broad midsection tapering at head and tail, dark stripe through eye, and females have double row of dark spots on side; Ningaloo Reef northwards; Indo-C. Pacific; to 7 cm.

219 BANDED PIPEFISH
Doryrhamphus dactyliophorus (Bleeker)
Inhabits coral reef crevices; distinguished by prominent light and dark bands; offshore reefs of North West Shelf; Indo-W. Pacific; to 18 cm.

220 JANSS'S PIPEFISH
Doryhamphus janssi (Herald & Randall)
Inhabits coral reef crevices; distinguished by red central section of body grading to blue on rear part and fan-shaped dark tail with pale centre and pale outer margin; Dampier Archipelago northwards; mainly W. Pacific; to 13 cm.

221 LADDER PIPEFISH
Festucalex scalaris (Günther)
Inhabits trawling grounds, amongst weeds; distinguished by short snout, and variegated pattern of light and dark spots, blotches and bars; Kalbarri to Ningaloo Reef; W.A. only; to 18 cm.

222 TIGER PIPEFISH
Filicampus tigris (Castelnau)
Inhabits sand-weed areas; distinguished by diagonal dark stripes on head, diffuse dark bars, and abruptly white belly; Shark Bay northwards; Australia only; to 35 cm.

223 TASSLED PIPEFISH
Halicampus brocki (Herald)
Inhabits coral and rocky reefs; distinguished by skin flaps and branched tassles on head and body, colour ranges from red to brown with pale spots and blotches; Abrolhos northwards; mainly W. Pacific; to 11 cm.

224 SHORT-NOSED PIPEFISH
Halicampus spinirostris (Dawson & Allen)
Inhabits coral and rocky reefs; distinguished by short snout, no skin flaps, and broad dark bars with narrow pale bars between them; Ningaloo Reef northwards; E. Indian Ocean and W. Pacific; to 11 cm.

225 RIBBONED PIPEFISH
Haliichthys taeniophorus Gray
Inhabits trawling grounds; distinguished by large size, elongate snout, bony knobs above eye, and prominent spines or knobs on body ridges; Shark Bay northwards; N. Australia and New Guinea; to 30 cm.

226 MANGROVE PIPEFISH
Hippichthys penicillus (Cantor)
Inhabits mangrove estuaries and lower reaches of freshwater streams; distinguished by relatively long snout, diagonal pale-edged stripe behind eye, and small pale spots on front part of body; Onslow northwards; N. Indian Ocean and W. Pacific; to 18 cm.

227 PALLID PIPEFISH
Solegnathus hardwickii (Gray)
Inhabits trawling grounds; distinguished by large size, pale colouration and dark marks along edge of back; Onslow northwards; mainly W. Pacific; to 50 cm.

228 WHITE-SADDLED PIPEFISH
Micrognathus micronotopterus (Fowler)
Inhabits inshore reefs and tide pools; distinguished by short snout, small, usually unbranched skin flaps on head and body, and 10-12 pale saddles on back; North West Cape northwards; Indo-Australian Archipelago; to 7 cm.

229 DOUBLE-ENDED PIPEFISH
Syngnathoides biaculeatus (Bloch)
Inhabits coastal waters, amongst weeds; distinguished by large size with deep, laterally compressed snout and prehensile tail; Shark Bay northwards; Indo-W. Pacific; to 29 cm.

230 SHORT-TAILED PIPEFISH
Trachyrhamphus bicoarctatus (Bleeker)
Inhabits sand, rubble or weed bottoms; distinguished by long thin tapering body, and tiny tail; Shark Bay northwards; Indo-W. Pacific; to 40 cm.

231 SLENDER PIPEFISH
Trachyrhamphus longirostris Kaup
Inhabits trawling grounds; similar to 230, but has thicker snout and fewer rings or body segments (41-53 behind anus versus 55-63); Exmouth Gulf northwards; Indo-W. Pacific; to 32 cm.

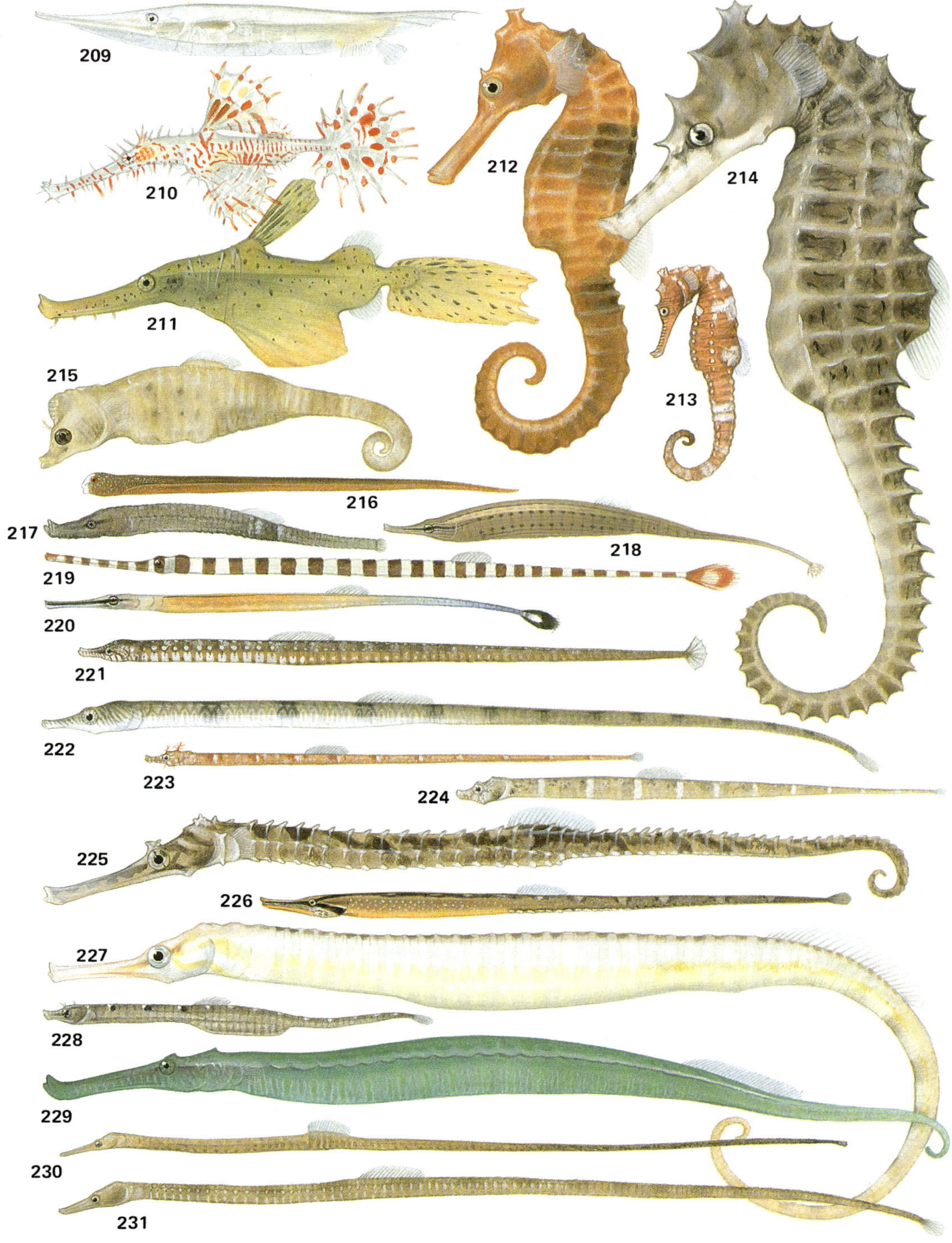

232 DWARF LIONFISH
Dendrochirus brachypterus (Cuvier)
Inhabits coral reefs; distinguished by large pectoral fins without elongate, free filamantous rays, similar to 233, but has more prominent curved bands on pectoral fins and bars on side are less well defined; also known as Short-finned scorpionfish; Abrolhos northwards; Indo-C. Pacific; to 15 cm.

233 ZEBRA LIONFISH
Dendrochirus zebra (Cuvier)
Inhabits coral reefs; distinguished by large pectoral fins without elongate, free filamentous rays, similar to 232, but bands on pectoral fins poorly defined and bars on side more distinct; also known as Butterfly scorpionfish; Shark Bay northwards; Indo-C. Pacific; to 18 cm.

234 RAGGED-FINNED FIREFISH
Pterois antennata (Bloch)
Inhabits coral reefs, usually in caves and crevices; distinguished by white filamentous pectoral rays, row of large dark spots at base of this fin and relatively few dark bars on head; also known as Spotfin lionfish; Carvarvon northwards; Indo-C. Pacific; to 30 cm.

235 DEEPWATER FIREFISH
Pterois mombasae (Smith)
Inhabits offshore reefs, usually below 40 m depth; distinguished by relatively short filamentous tips on pectoral fin which is densely spotted and maze of dark and light bands on tail base; Onslow northwards; Indo-W. Pacific; to 16 cm.

236 SPOTLESS FIREFISH
Pterois russelli Bennett
Inhabits offshore reefs, usually below 20-30 m depth; distinguished by filamentous pectoral rays and lack of spots on dorsal, anal, and tail fins; Exmouth Gulf northwards; Indo-W. Pacific; to 30 cm.

237 RED FIREFISH
Pterois volitans (Linnaeus)
Inhabits coral reefs and southern rocky reefs, usually in caves or crevices; distinguished by broad filamentous pectoral rays, similar to 236, but has spots on dorsal, anal, and tail fins; also known as Butterfly cod and Volitans lionfish; Fremantle northwards; Indo-C. Pacific; to 38 cm.

238 ROUNDFACE FIREFISH
Brachypterois serrulatus (Richardson)
Inhabits deeper trawling grounds; distinguished by dusky fan-like pectoral fins without free filamentous rays, lack of distinct bars on side, and faint spotting on dorsal, anal, and tail fins; Derby northwards; E. Indian Ocean and W. Pacific; to 10 cm.

239 ESTUARINE STONEFISH
Synanceja horrida (Linnaeus)
Inhabits sand or mud bottoms amongst rocks, sometimes under dead coral slabs; distinguished by stone-like appearance and warty projections on body; fin spines extremely venemous; Shark Bay northwards; E. Indian Ocean and W. Pacific; to 47 cm.

240 MONKEYFISH
Erosa erosa Langsdorf)
Inhabits trawling grounds; similar to 241, but has slight hump in front of dorsal fin, white spots on outer part of pectoral fin and narrow cross-bars on tail; also known as Pitted scorpionfish; Port Hedland northwards; mainly W. Pacific; to 15 cm.

241 DAMPIER STONEFISH
Dampierosa daruma Whitley
Inhabits coastal waters in the vicinity of reefs, sometimes under wharves; similar to 240, but has rounded head profile without hump in front of dorsal fin, broad white band (instead of spots) on outer pectoral fin and a single dark bar across middle part of tail; Exmouth Gulf northwards; N.W. Australia only; to 13 cm.

242 COCKATOO WASPFISH
Ablabys taenianotus (Cuvier)
Inhabits coastal reefs; distinguished by thin, laterally compressed body, vertical profile of snout, and elevated rays at front of dorsal fin; Ningaloo Reef northwards; Indian Ocean and Indo-Australian Archipelago; to 10 cm.

243 BLACKSPOT WASPFISH
Liocranium praepositum Ogilby
Inhabits trawling grounds; similar to 242, but has larger eye, lacks elevated rays at front of dorsal fin, and has prominent dark blotch above pectoral fin; Exmouth Gulf northwards; N. Australia only; to 13 cm.

244 PLUMB-STRIPED STINGFISH
Minous versicolor Ogilby
Inhabits trawling grounds; distinguished by irregular stripes and blotches on upper side, and wavy cross bands on dorsal and tail fins, also has free lowermost pectoral ray; Shark Bay northwards; N. Australia only; to 11 cm.

245 SPOTFIN WASPFISH
Paracentropogon vespa Ogilby
Inhabits trawling grounds; similar to 242 and 243 but has large dark blotch on base of front part of dorsal fin; Shark Bay northwards; N. Australia only; to 9 cm.

246 DEMON STINGER
Inimicus didactylus (Pallas)
Inhabits rubble bottoms, frequently in the vicinity of coral reefs; distinguished by large upturned mouth, free pair of rays at lowermost part of pectoral fin, and prominent dorsal spines, similar to 247, but has yellow tail with dark submarginal bar and pale band across pectoral fin; Shark Bay northwards; Indo-Australian Archipelago; to 18 cm.

247 SPOTTED STINGER
Inimicus sinensis (Valenciennes)
Inhabits rubble bottoms; similar to 246, but tail mainly dusky or spotted (not yellow with dark cross bar) and lacks pale band on pectoral fin; Abrolhos northwards; E. Indian Ocean and W. Pacific; to 25 cm.

WARNING! The fishes shown on Plates 15 and 16 possess venomous fin spines and handling of live or freshly dead specimens should be avoided. The Estuarine Stonefish (239) is amongst the most venomous of all fishes and is capable of causing death. Symptoms of scorpionfish stings range from a bee-sting type sensation to violent pain and may lead to unconsciousness or extended coma. Immersing the wound in very hot water is an effective first aid treatment and a physician should be consulted immediately.

248 SHORT-FINNED WASPFISH
Apistops coloundra (De Vis)
Inhabits trawling grounds; distinguished by 5 chin barbels, diffuse dark stripes on side and spot on dorsal fin, similar to 249, but pectoral fins shorter (do not reach the rear part of anal fin); Abrolhos northwards; N. Australia only; to 12 cm.

249 LONG-FINNED WASPFISH
Apistus carinatus (Bloch & Schneider)
Inhabits trawling grounds; similar to 248, but with 3 chin barbels, no stripes on side and pectoral fins usually longer, reaching to rear part of anal fin; Shark Bay northwards; Indo-W. Pacific; to 18 cm.

250 COD SCORPIONFISH
Peristrominous dolosus Whitley
Inhabits trawling grounds; distinguished by elongate brown body without distinct markings and pointed snout; Shark Bay northwards; N. Australia only; to 9 cm.

251 MARBLED STINGFISH
Cottapistus cottoides Cuvier
Inhabits deep offshore reefs and trawling grounds; distinguished by forward position (over eye) of dorsal fin origin, hump on snout, and very small scales; Monte Bello Islands northwards; mainly W. Pacific; to 12 cm.

252 NORTHERN SCORPIONFISH
Parascorpaena picta (Cuvier)
Inhabits crevices of coral and rocky reefs; distinguished by well camouflaged appearance with skin flaps and tentacles on head and body, 12 dorsal spines, and spine above upper jaw that curves forward; Shark Bay northwards; Indo-W. Pacific; to 16 cm.

253 GUAM SCORPIONFISH
Scorpaenodes guamensis (Quoy & Gaimard)
Inhabits coral reef crevices; distinguished by dark spot on upper part of gill cover and 13 dorsal spines; Point Quobba northwards; Indo-C. Pacific; to 12 cm.

254 ORNATE SCORPIONFISH
Scorpaenodes varipinnis Smith
Inhabits coral reef crevices; distinguished by red colour with white blotches on head and along middle of sides, a curved dark band across pectoral fin and 13 dorsal spines; Ningaloo Reef northwards; Indo-W. Pacific; to 7 cm.

255 LITTLE SCORPIONCOD
Scorpaenodes sp.
Inhabits coral reef crevices; distinguished by red or brown spots on fins, pale bar across tail base, lack of dark spot on gill cover, and 13 dorsal spines; possibly a colour variation of 256; to 8 cm.

256 PYGMY SCORPIONFISH
Scorpaenodes scaber (Ramsay & Ogilby)
Inhabits coastal and estuaries reefs; similar to 253, but has dark spot on lower edge of gill cover instead of upper part; Abrolhos northwards; Indo-W. Pacific; to 8 cm.

257 RAGGY SCORPIONFISH
Scorpaenopsis venosa (Cuvier)
Inhabits coral reef crevices; distinguished by skin flaps and tentacles on head and body; relatively tall dorsal fin, and 12 dorsal spines, juveniles usually more ornate as shown; Ningaloo Reef northwards; Indo-C. Pacific; to 18 cm.

258 FALSE STONEFISH
Scorpaenopsis diabolus (Cuvier)
Inhabits coral reef crevices and rubble bottoms; distinguished by well camouflaged appearance, humped back, and bright yellow-orange patch on inner surface (not shown) of pectoral fins; often confused with the Estuarine stonefish (239) which is far more venomous; Ningaloo Reef northwards; Indo-C. Pacific; to 18 cm.

259 WHITE-BELLIED ROUGEFISH
Tetraroge leucogaster (Richardson)
Inhabits trawling grounds; distinguished by laterally compressed body, forward position (above rear part of eye) of dorsal fin origin, relatively large eye, and complete lack of scales; Broome northwards; Indo-W. Pacific; to 8 cm.

260 LEAF SCORPIONFISH
Taenionotus triacanthus Lacepède
Inhabits offshore coral reefs; distinguished by thin leaf-like body and tall dorsal fin, several colour varieties encountered including ones that are predominantly reddish, yellow, or black; North West Shelf; Indo-C. Pacific; to 10 cm.

261 DEEPSEA SCORPIONFISH
Setarches guentheri Johnston
Inhabits offshore trawling grounds; distinguished by overall red colour without distinct marks, somewhat pointed snout, very stout spines on edge of cheek, and 12 dorsal spines, Broome northwards; worldwide in tropical seas; to 23 cm.

SCORPIONFISH COLOURS

The coloration of many of the scorpionfishes shown on this plate is extremely variable, depending on size, depth, and habitat. It is not unusual for the same species to exhibit very different patterns at a particular locality. For example the northern scorpionfish (252) is often mottled brown when found among rocks in shallow weedy areas, and red (as shown) if seen in caves on deeper sections of the reef.

SCORPIONFISHES AND ALLIED FAMILIES

The families featured on Plates 15-17 are members of the order Scorpaeniformes. They are distinguished by a bony ridge on the cheek and the head is frequently spiny or tassled. All are bottom living fishes that occur in a variety of depths and habitats. They often exhibit variegated colour patterns that blend well with their surroundings, thus enabling them to remain undetected by small fishes, crustaceans and other organisms upon which they feed. Scorpionfishes (family Scorpaenidae: 232-261) are found in all tropical and temperate seas and are commercially important in some areas.

262 SANDPAPER VELVETFISH
Adventor elongatus (Whitley)
Inhabits trawling grounds; distinguished by elongate body shape and bony knobs on head, similar to 266 but dorsal fin begins further back on head; Exmouth Gulf northwards; Australia only; to 11 cm.

263 DUSKY VELVETFISH
Aploactis aspersa (Richardson)
Inhabits trawling grounds; has longer anal fin base than other velvet fishes on this page; colour sometimes brown; Shark Bay northwards; E. Indian Ocean and W. Pacific; to 9 cm.

264 THIN VELVETFISH
Coccotropus sp.
Inhabits trawling grounds and sandy areas near reefs; distinguished by laterally compressed body, steep forehead, and tall anterior part of dorsal fin that begins above eye; similar to some scorpionfishes (232-261) but has bony knobs (versus spines) on head and body covered with prickles; Shark Bay northwards; possibly Australia only; to 5 cm.

265 THREEFIN VELVETFISH
Neoaploactis tridorsalis Eschmeyer & Allen
Inhabits sand or rubble bottoms near reefs; distinguished by 3 separate dorsal fins; Rottnest Island to Shark Bay; Australia only; to 5 cm.

266 BEARDED VELVETFISH
Paraploactis intonsa Poss & Eschmeyer
Inhabits trawling grounds; similar to 262 and 263, but deeper bodied, steeper forehead, and dorsal fin begins above eye; *P. pulvinus* (not shown) is similar, but lacks prickles on ventral surface of lower jaw; Shark Bay; W.A. only; to 14 cm.

267 DARK-FINNED VELVETFISH
Erisphex aniarus (Thompson)
Inhabits trawling grounds, distinguished by pale colouration except for blackish fins; Broome northwards; E. Indian Ocean and W. Pacific; to 10 cm.

268 LONG-FINNED GURNARD
Lepidotrigla argus Ogilby
Inhabits trawling grounds; distinguished by pair of short forward projecting spines on snout and enlarged fan-like pectoral fins with blue and yellow markings and blue-edged black spot; Dampier northwards; Australia only; to 18 cm.

269 BLACK-FINNED GURNARD
Pterygotrigla leptacanthus (Günther)
Inhabits trawling grounds; similar to 268, but lacks scales, has longer forward projecting spines on snout, and blackish pectoral fins; Derby northwards; Indo-Australian Archipelago; to 15 cm.

270 HALF-SPOTTED GURNARD
Pterygotrigla hemisticta (Temminck & Schlegel)
Inhabits trawling grounds; similar to 269, but has black spot on first dorsal fin and scattered brown spots on back; Derby northwards; E. Indian Ocean and W. Pacific; to 25 cm.

271 SLENDER ARMOURED-GURNARD
Peristedion liorhynchus (Günther)
Inhabits trawling grounds; distinguished by dark margin on dorsal fins and banded pectoral fins, shape from dorsal view similar to 272; Port Hedland northwards; E. Indian Ocean and W. Pacific; to 25 cm.

272 SPOTTED ARMOURED-GURNARD
Satyrichthys rieffeli (Kaup)
Inhabits trawling grounds; distinguished by black spotting on head, body and dorsal fin; shape from side view similar to 271, Derby northwards; E. Indian Ocean and W. Pacific; to 20 cm.

273 FRINGE-EYED FLATHEAD
Cymbacephalus nematophthalmus (Günther)
Inhabits sand bottoms; distinguished by 6-9 skin tentacles above eye (versus 0-1 in most flatheads), 7-8 dusky bands across nape and back, extending on to sides and strongly variegated pattern on fins; Shark Bay northwards; to 58 cm; .655 kg.

274 DWARF FLATHEAD
Elates ransonnetti (Steindachner)
Inhabits sand bottoms; distinguished by 6 dorsal spines (versus 7-9 spines for other flatheads), filamentous upper lobe of tail, and semi-transparent appearance; Broome northwards; Indo-Australian Archipelago; to 19 cm.

275 HARRIS'S FLATHEAD
Inegocia harrisii (McCulloch)
Inhabits sand bottoms; distinguished by overall orange-brown colour with fine brown spots on back and white below, irregular dark bars on pectoral fins, and elongate dark streaks on tail; Carnarvon northwards; Australia only; to 20 cm.

276 SPINY FLATHEAD
Onigocia spinosa (Temminck & Schlegel)
Inhabits sand bottoms; distinguished by numerous small spines on head which is very broad (when viewed from above), outer half of spiny dorsal fin dark brown; irregular brown bars across back and sides, and largely blackish pelvic fins with yellowish tips; Cockburn Sound northwards; E. Indian Ocean and W. Pacific; to 9 cm.

277 NORTHERN SAND FLATHEAD
Platycephalus arenarius Ramsay & Ogilby
Inhabits sand bottoms; distinguished by black stripes on tail; Shark Bay northwards; Australia only; to 45 cm; 4.99 kg. ★★★

278 BAR-TAILED FLATHEAD
Platycephalus endrachtensis Quoy & Gaimard
Inhabits sand bottoms; distinguished by black stripes on tail, similar to 277 but has fewer black stripes and a yellow blotch on upper part of tail; Fremantle northwards; Australia only; to 26 cm; 2.8 kg. ★★★

279 RUSTY FLATHEAD
Suggrundus japonica (Tilesius)
Inhabits sand bottoms; a reddish-brown flathead similar to 275 in general appearance, but has definite dark spots on tail (versus elongate streaks); sometimes referred to as *S. isacathus;* Fremantle northwards; Indo-W. Pacific; to 20 cm. ★★★

280 HEART-HEADED FLATHEAD
Sorsogona tuberculata (Cuvier)
Inhabits sand bottoms; distinguished by prominent black area on outer part of pectoral fins and strongly barred pelvic fins; Shark Bay northwards; Indo-W. Pacific; to 50 cm. ★★★

281 OLIVE-TAILED FLATHEAD
Rogadius asper (Cuvier)
Inhabits sand bottoms; distinguished from other flatheads by forward directed spine on lower edge of cheek, similar in colour to 280, but has broad dusky margin on spiny dorsal fin and lacks faint spotting on tail; Broome northwards; Indo-W. Pacific; to 17 cm. ★★★

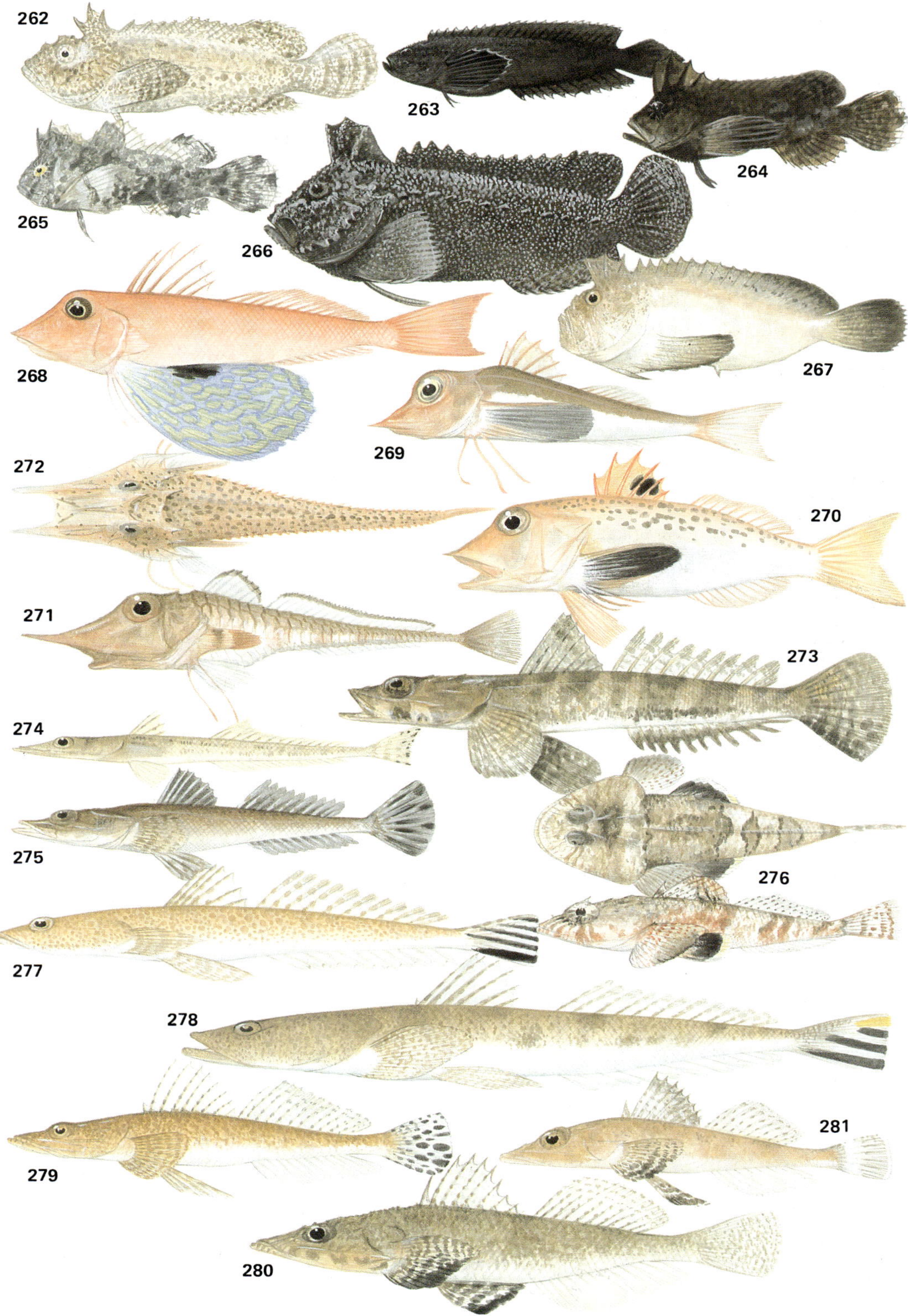

282 WHITE-LINED ROCKCOD
Anyperodon leucogrammicus (Valenciennes)
Inhabits coral reefs; distinguished by elongate shape and pattern of pale longitudinal stripes and numerous dark spots; Dampier Archipelago northwards; Indo-W. Pacific; to 50 cm; 1.259 kg. ★★★

283 FALSE SCORPIONFISH
Centrogenys vaigiensis (Quoy & Gaimard)
Inhabits sand and weed flats, frequently around small rocky outcrops; resembles members of scorpionfish family (232-261), but lacks numerous head spines and is *not* venomous (fin spines); Shark Bay northwards; Indo-Australian Archipelago; to 15 cm.

284 BROWN-BANDED ROCKCOD
Cephalopholis boenack Bloch
Inhabits dead reefs in protected inshore waters; distinguished by overall brown colour with faint dark bars on sides and large blackish spot at rear of gill cover; formerly known as *C. pachycentron;* Ningaloo Reef northwards; Indo-W. Pacific; to 22 cm. ★★

285 PEACOCK ROCKCOD
Cephalopholis argus Schneider
Inhabits caves and crevices of coral reefs; distinguished by numerous dark-edged blue spots on head, body and fins, whitish area in front of pectoral fin and 5-6 pale bars on posterior part of body frequently present; Abrolhos northwards; Indo-C. Pacific; to 50 cm. ★★★

286 CORAL COD
Cephalopholis miniata (Forsskål)
Inhabits caves and crevices of coral reefs; distinguished by numerous blue spots on orange-red to red-brown background; also known as Coral trout; Abrolhos northwards; Indo-W. Pacific; to 41 cm; 1.6 kg. ★★★

287 BLUE-LINED ROCKCOD
Cephalopholis formosa (Shaw)
Inhabits inshore coral reefs; distinguished by narrow dark blue stripes on head, body, and fins; Dampier Archipelago northwards (rare); Indo-W. Pacific; to 33 cm. ★★★

288 RED-SPOTTED ROCKCOD
Cephalopholis leopardus (Lacepède)
Inhabits caves and crevices of coral reefs; distinguished by oblique dark streaks on upper and lower lobe of tail, dark patch on upper edge of tail base, and dark spot at rear of gill cover; North West Shelf; Indo-C. Pacific; to 20 cm. ★★

289 TOMATO ROCKCOD
Cephalopholis sonnerati (Valenciennes)
Inhabits coral and rock reefs, adults often in deeper water (30-100 m); distinguished by orange-red to reddish brown colour, frequently with scattered whitish blotches; a brown variety is shown on Plate 19 (308); juveniles are pale pinkish (see opposite Plate); Cape Cuvier northwards; Indo-W. Pacific; to 58 cm; 2.2 kg. ★★★

290 FLAG-TAILED ROCKCOD
Cephalopholis urodeta (Schneider)
Inhabits coral reef crevices; distinguished by pale oblique streaks on lobes of tail; North West Shelf; Indo-W. Pacific; to 23 cm. ★★

291 BARRAMUNDI COD
Cromileptes altivelis (Valenciennes)
Inhabits caves and crevices of coral reefs; distinguished by small head, laterally compressed body and polkadot pattern; Dirk Hartog Island northwards; Indo-W. Pacific; to 70 cm; 4.8 kg. ★★★★

292 COOPER'S FAIRY BASSLET
Anthias cooperi Regan
Inhabits offshore reefs to at least 160 m depth; male (shown here) has red spot on sides and elongate filaments on pelvic, anal, and tail fins; female lacks these features and has a red spot on the tip of each lobe of the tail; Barrow Island northwards; Indo-W. Pacific; to 13 cm.

293 PEARL-SPOTTED FAIRY BASSLET
Selenanthias analis Tanaka
Inhabits offshore trawling grounds; distinguished by pearly spots on sides and black spot on anal fin; North West Shelf; Northern Australia and Japan to Taiwan; to 16 cm.

294 CITRON PERCHLET
Plectranthias megalophthalmus Fourmanoir & Randall
Inhabits deep offshore trawling grounds to 360 m; distinguished by yellow colouration; North West Shelf; New Caledonia and Australia; to 8 cm.

295 SPOTTED PERCHLET
Plectranthias wheeleri Randall
Inhabits deep offshore trawling ground to 230 m; distinguished by irregular blotches on sides; North West Shelf; Australia and Indonesia; to 10 cm.

296 JAPANESE PERCHLET
Plectranthias japonicus (Steindachner)
Inhabits deep trawling grounds to at least 200 m; distinguished by red-orange colouration, sometimes with darker blotches on sides; North West Shelf; Australia and Japan; to 15 cm.

BASSLETS, PERCHLETS, ETC.

The Rockcods, Coral Cod, Basslets, and Perchlets featured on this plate are members of the family Serranidae, a group that is well represented in all tropical and subtropical seas. The Basslets (292-293, and 325-326 on Plate 21) and Perchlets (294-296) are included in the subfamily Anthiinae. Most of the approximately 100 species in this group are small (usually under 20 cm) fishes that exhibit a broad spectrum of vivid colours, particularly shades of red, purple, and yellow. Huge shoals of colourful basslets are a common sight adjacent to steep outer reef slopes throughout the Indo-west Pacific region. They occur in both shallow water and to depths of at least 200 m. Like the gropers, cods, and coral trouts, they are hermaphroditic, with adult females capable of transformation to the male sex. Many species have a harem-type social structure in which a single dominant male is associated with several females. The male and female colour patterns may be very different.

297 RED-FLUSHED ROCKCOD
Aethaloperca rogaa (Forsskål)
Inhabits coral reefs in the vicinity of caves; distinguished by elevated shape of body, dark colouration and white edge on tail; North West Cape; Indo-W. Pacific; to 60 cm; 2.5 kg. ★★★

298 BLUNT-HEADED ROCKCOD
Epinephelus amblycephalus (Bleeker)
Inhabits deeper offshore reefs; distinguished by 5 dark bars on body; Shark Bay northwards; mainly W. Pacific; to 45 cm. ★★★

299 YELLOW-SPOTTED ROCKCOD
Epinephelus areolatus (Forsskål)
Inhabits inshore reefs, usually around small coral heads in sandy areas or amongst sea grass; distinguished by dense pattern of large round spots and truncate (not rounded) tail; Ningaloo Reef northwards; Indo-W. Pacific; to 35 cm; .75 kg. ★★★

300 OCELLATED ROCKCOD
Epinephelus caeruleopunctatus (Bloch)
Inhabits coral reefs, near caves and crevices; similar to 313, but spots more uniformly round and smaller; Onslow northwards; Indo-W. Pacific; to 60 cm. ★★★

301 CORAL ROCKCOD
Epinephelus corallicola (Valenciennes)
Inhabits shallow, silty reefs and estuaries; distinguished by round black spots on grey background; Port Hedland northwards; W. Pacific; to 31 cm. ★★★

302 BLACK-TIPPED COD
Epinephelus fasciatus (Forsskål)
Inhabits coral reefs and rocky bottoms to 100 m depth; distinguished by reddish bars and narrow black border on front part of dorsal fin; Lancelin northwards; Indo-C. Pacific; to 40 cm; .51 kg. ★★★

303 SPOTFIN ROCKCOD
Epinephelus latifasciatus (Temminck & Schlegel)
Inhabits sand and rock bottoms on the continental shelf between 20-200 m depth; distinguished by thin dark lines (or sometimes rows of faint spots) on sides, and spots on dorsal fin and tail; North West Shelf; Indo-W. Pacific; to 70 cm. ★★★

304 FLOWERY COD
Epinephelus fuscoguttatus (Forsskål)
Inhabits coral reefs and rocky bottoms; similar to 307, but irregular brown blotches on sides generally more diffuse and spot at upper base of tail smaller; best means of separation is pectoral fin ray count (18-20, usually 19); Dampier Archipelago northwards; Indo-W. Pacific; to 90 cm; 2.75 kg. ★★★

305 THREE-LINED ROCKCOD
Epinephelus heniochus Fowler
Inhabits offshore trawling grounds to at least 80 m depth; distinguished by overall pinkish-red colour and narrow stripes (often very faint) on head; North West Shelf; mainly W. Pacific; to 30 cm. ★★★

306 HONEYCOMB COD
Epinephelus merra Bloch
Inhabits protected inshore coral reefs; distinguished by dense network of large spots on body and fins; a similar species *E. tauvina* (not shown) has a more slender shape and more space between spots; Dampier Archipelago northwards; Indo-C. Pacific; to 28 cm. ★★★

307 SMALL-TOOTHED COD
Epinephelus microdon (Bleeker)
Inhabits coral reefs, often around large bommies; similar to 304 but has more distinct oblique bands on back and head and fewer pectoral rays (16-17, usually 17); North West Shelf; Indo-C. Pacific; to 61 cm. ★★★

308 TOMATO ROCKCOD
Cephalopholis sonnerati (Valenciennes)
Inhabits coral reefs in the vicinity of caves and crevices; the red variety and juvenile are shown on Plate 18 (289); brown variety shown here is sometimes seen at Ningaloo Reef in 12-20 m depth; Cape Cuvier northwards; Indo-W. Pacific; to 58 cm. ★★★

309 LONG-FINNED ROCKCOD
Epinephelus quoyanus (Valenciennes)
Inhabits silty inshore reefs; similar to 306 but spotting is less dense and has oblique dark bands just below and slightly in front of pectoral fin base; Onslow northwards; mainly W. Pacific; to 35 cm; 2.268 kg. ★★★

310 CHINAMAN ROCKCOD
Epinephelus rivulatus (Valenciennes)
Inhabits inshore coral and rock reefs, usually around small coral heads or amongst weed; distinguished by oblique brown bars on sides and white blotches on heads; Rottnest Island northwards; Indo-W. Pacific; to 35 cm; 1.33 kg. ★★★

311 RADIANT ROCKCOD
Epinephelus radiatus (Day)
Inhabits sand and rock bottoms, usually trawled between 80-160 m depth; distinguished by broad diagonal bands on head and body; North West Shelf; Indo-W. Pacific; to 70 cm. ★★★

297
298
299
300
301
302
303
303 juv.
304
305
306
307
308
309
310
311

312 FROSTBACK COD
Epinephelus bilobatus Randall & Allen
Inhabits inshore coral reefs, usually where there is some sand; distinguished by white area on upper back and dark spots along base of dorsal fin; Point Quobba northwards; north-western Australia; to 40 cm. ★★★

313 WHITE-BLOTCHED ROCKCOD
Epinephelus multinotatus (Peters)
Inhabits inshore coral reefs and deeper offshore trawling grounds; distinguished by irregular white blotches; also known as Rankin's rockcod; Abrolhos northwards; Indian Ocean; to 100 cm; 9.07 kg. ★★★

314 SIX BANDED ROCKCOD
Epinephelus sexfasciatus (Valenciennes)
Inhabits offshore trawling grounds to 70 m depth; distinguished by combination of bars on sides and spotted tail; Exmouth Gulf northwards; south-east Asia and Indo-Australian Archipelago; to 26 cm. ★★★

315 ESTUARY COD
Epinephelus suillus (Valenciennes)
Inhabits inshore coral reefs and estuaries; distinguished by oblique bands on sides overlaid with red-brown spotting; frequently misidentified as *E. malabaricus* or *E. tauvina;* also known as Greasy cod; Rottnest Island northwards; Indo-W. Pacific; to at least 180 cm; 106.14 kg. ★★★

316 POTATO COD
Epinephelus tukula Morgans
Inhabits coral reefs in the vicinity of caves and crevices; distinguished by large ovate spots; more common on offshore reefs such as Rowley Shoals; Ningaloo Reef northwards; Indo-W. Pacific; to 140 cm; 6.237 kg. ★★★

317 WOORE'S ROCKCOD
Tristotropis dermopterus (Temminck & Schlegel)
Inhabits offshore trawling grounds; distinguished by 11 dorsal spines, rounded tail, 2 short canine teeth on each side at front of jaws, and ctenoid scales (i.e. rough to the touch); Dampier Archipelago northwards; Australia and Japan to Taiwan; to 45 cm. ★★★

318 CORAL TROUT
Plectropomus leopardus (Lacepède)
Inhabits coral reefs; distinguished by numerous small round spots on head and body; also known as Leopard cod; Dongara northwards; mainly W. Pacific; to 75 cm; 15.5 kg. ★★★★

319 POLKADOT COD
Plectropomus areolatus (Rüppell)
Inhabits coral reefs; distinguished by numerous dark-edged round spots on head and body; North West Shelf; Indo-W. Pacific; to 70 cm. ★★★

320 BAR-CHEEKED CORAL TROUT
Plectropomus maculatus (Bloch)
Inhabits coral reefs; similar to 318, but has fewer, more widely-spaced spots and those on head are elongate in shape; also known as Coral cod; Abrolhos northwards; south-east Asia and Indo-Australian Archipelago; to 70 cm; 5.95 kg. ★★★

321 VERMICULAR COD
Plectropomus oligocanthus (Bleeker)
Inhabits offshore coral reefs; distinguished by bright pinkish-red colour, blue spots (some of which are elongated) on body and fins, and relatively tall dorsal (posterior half) and anal fins; North West Shelf; mainly W. Pacific; to 56 cm. ★★★

322 CORONATION TROUT
Variola louti (Forsskål)
Inhabits inshore coral reefs and deeper offshore reefs to 100 m; distinguished by bright colour pattern and distinct lunar-shaped tail; Shark Bay northwards; Indo-W. Pacific; known to cause ciguatera poisoning in some areas outside of W.A.; to 80 cm; 3.0 kg. ★★★

323 QUEENSLAND GROPER
Epinephelus lanceolatus (Bloch)
Inhabits coral reefs and rocky areas, often in the vicinity of caves; mainly distinguished by its huge size, is one of the largest of all bony fishes; placed in the genus *Promicrops* by some authors; Rottnest Island northwards; Indo-W. Pacific; to 270 cm; 150.45 kg. ★★★

> ## GIANTS OF THE REEF
> Although most of the serranid fishes featured on Plates 18-21 are well under 1 metre in length at maximum size, three of the species featured on this plate are giants by comparison. Foremost in this respect is the Queensland Groper (323) which attains a size of at least 2-3 metres and weight of about 400 kg. Although the common name suggests a very limited distribution it is actually found over a huge area of the tropical Indo-Pacific extending from East Africa to the islands of the central Pacific Ocean. This fish is generally harmless, but because of its great size it should be treated with respect, particularly when encountered while diving. They have a curious nature and will often approach a diver at close range. Some years ago a diver on the southern Great Barrier Reef foolishly offered a feeding of fish to one of these behemouths. The Giant Groper engulfed the fish along with the diver's arm up to the elbow. Although the teeth are very small, there are lots of them and the diver suffered severe lacerations when jerking his arm out of the mouth. Although not exactly in the same league as the Giant Groper, other large species that grow over 1 metre include the Estuary Cod (314) which is common along the coast north of Exmouth Gulf and the Potato Cod (316), thus far known in W.A. only at Rowley Shoals.

312
313
314
315
316
317
318
319
320
321
322
322 juv.
323
323 juv.

324 YELLOW EMPEROR
Diploprion bifasciatum Cuvier
Inhabits coral reefs; distinguished by bar through eye and broad dark bar across middle of body; also known as Two-banded perch; Ningaloo Reef northwards; N. Indian Ocean and W. Pacific; to 38 cm.

325 DEEPSEA FAIRY BASSLET
Anthias rubrizonatus Randall
Inhabits offshore reefs, usually in deeper water (over 50 m depth); distinguished by dark saddle below dorsal spines and pale diagonal stripe across lower part of head; Barrow Island northwards; Indo-Australian Archipelago; to 12 cm.

326 LITTLE FAIRY BASSLET
Sacura parva Heemstra & Randall
Inhabits offshore reefs, usually in deeper water (over 50 m depth); distinguished by elongate third dorsal spine and filament at front of soft dorsal fin; Barrow Island northwards; Indo-Australian Archipelago; to 12 cm.

327 RAINFORD'S PERCH
Rainfordia opercularis McCulloch
Inhabits coral reefs, usually in caves; distinguished by elongate body, flattened head and stripe pattern; has mucus that is toxic to other fishes; Dampier Archipelago northwards; N. Australia only; to 15 cm.

328 SIX-LINED PERCH
Grammistes sexlineatus (Thunberg)
Inhabits coral reefs, usually in caves or crevices; distinguished by narrow yellow stripes on side; has mucus that is toxic to other fishes; Ningaloo Reef northwards; Indo-C. Pacific; to 27 cm.

329 FALSE GRAMMA
Pseudogramma polyacantha (Bleeker)
Inhabits coral reefs, in caves and crevices; distinguished by small size, and dusky brown pattern with faint blotches; Ningaloo Reef northwards; Indo-C. Pacific; to 7 cm.

330 BANDED LONGFIN
Belonepterygium fasciolatum (Ogilby)
Inhabits coral reefs, in caves and crevices; distinguished by dark stripe through eye, narrow dark bars on side, and long thread-like pelvic fins; Abrolhos northwards; N. Australia only; to 5 cm.

331 COMET
Calloplesiops altivelis (Steindachner)
Inhabits coral reefs, in caves and crevices; distinguished by white spots on head, body, and fins, and pale-edged black spot at rear of dorsal fin; Ningaloo Reef northwards; Indo-W. Pacific; to 16 cm.

332 RED-TIPPED LONGFIN
Plesiops coeruleolineatus Rüppell
Inhabits coral reefs, in caves and crevices; distinguished by deeply incised dorsal fin profile and red margin on dorsal fin; Ningaloo Reef northwards; Indo-W. Pacific; to 8 cm.

333 LINED DOTTYBACK
Labracinus lineatus (Castelnau)
Inhabits coral reefs and rocky areas, usually in crevices; distinguished by relatively large size and numerous narrow stripes on body and fins; also known as Lined cichlops; Jurien Bay northwards; E. Indian Ocean and W. Pacific; to 25 cm; .125 kg.

334 BROWN DOTTYBACK
Pseudochromis fuscus (Müller & Troschel)
Inhabits the vicinity of coral reefs; has two distinct colour phases, one entirely yellow and the other dusky brown or purplish; Ningaloo Reef northwards; E. Indian Ocean and W. Pacific; to 8 cm.

335 MARSHALL DOTTYBACK
Pseudochromis marshallensis Schultz
Inhabits coral reefs; distinguished by scales on sides having orange centres; Ningaloo Reef northwards; mainly W. Pacific; to 8 cm.

336 LONGFIN DOTTYBACK
Pseudochromis punctatus (Richardson)
Inhabits sandy areas with occasional coral or rocky outcrops; distinguished by dark bluish-brown colour and relatively tall dorsal and anal fins; also the dorsal fin begins farther forward than in similar species; Shark Bay northwards; N.W. Australia only; to 7.5 cm.

337 SPOTTED DOTTYBACK
Pseudochromis quinquedentatus McCulloch
Inhabits sand or rubble areas with occasional outcrops that serve as shelter; distinguished by small dark spots arranged in longitudinal rows on side; Exmouth Gulf northwards; N. Australia only; to 9 cm.

338 YELLOWHEAD DOTTYBACK
Pseudochromis tapeinosoma Bleeker
Inhabits coral reef crevices; male distinguished by yellowish area on head and breast with broad pale margin on tail and dorsal fin, females by yellow-edged red tail; Ningaloo Reef northwards; E. Indian Ocean and W. Pacific; to 7 cm.

339 YELLOWFIN DOTTYBACK
Pseudochromis wilsoni Whitley
Inhabits coral reef crevices and rubble areas; males distinguished by plain purplish colour and yellow-orange iris, females by yellowish dorsal fin and yellow edges on tail; Shark Bay northwards; N. Australia only; to 8 cm.

340 ROSE DEVILFISH
Pseudoplesiops rosae Schultz
Inhabits coral reef crevices; distinguished by small size, yellow-brown colour, large eye, and thread-like pelvic fins; Ningaloo Reef northwards; Indo-C. Pacific; to 4 cm.

SLIMY SOAPFISHES

Most of the species illustrated on this plate are cryptic dwellers of caves and crevices of the reef. The soapfishes (324, 327-328) of the family Grammistidae obtain their name from a slimy exterior mucus coat. It is extremely bitter to the taste and probably offers some measure of protection from larger predatory species. Several years ago we experimentally offered a live specimen of the Six-lined Perch (328) to a hungry Red Firefish (237), a voracious predator of small fishes and invertebrates. The Firefish stalked the Perch around the confines of the aquarium, then suddenly engulfed the front half of the fish in its mouth. The prey was spat out faster than it was taken in and escaped unharmed thanks to its unpalatable slime.

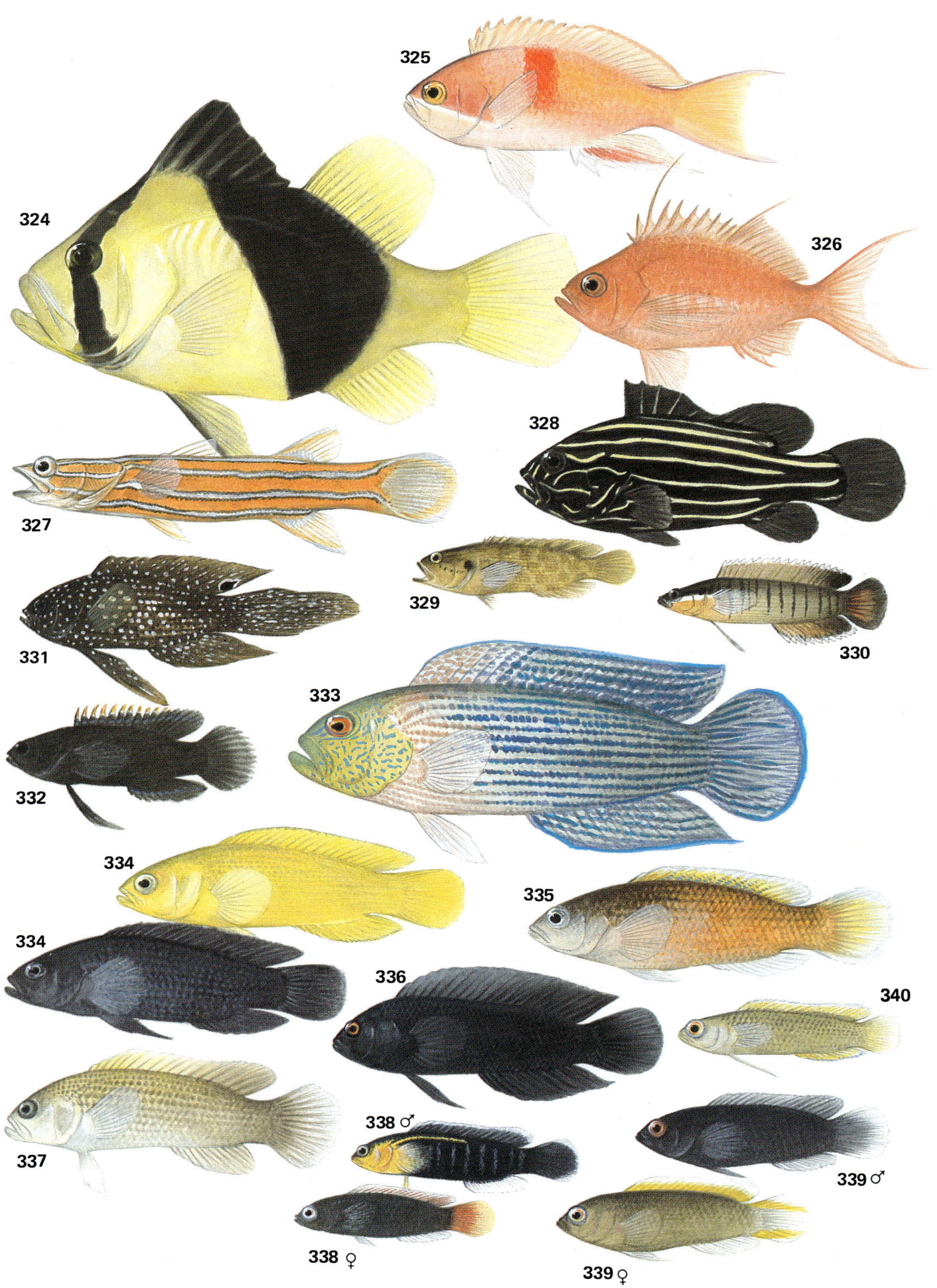

PLATE 22: PERCHLETS, GRUNTERS, AND WHITINGS

341 BARRAMUNDI
Lates calcarifer (Bloch)
Inhabits coastal waters, entering estuaries and fresh water; highly esteemed as a food fish and for its angling qualities; distinguished by silvery appearance, deeply notched dorsal fin, 'humped' back, and glassy eye; Onslow northwards; N. Indian Ocean and W. Pacific; to 150 cm; 15.4 kg. ★★★★

342 SAND BASS
Psammoperca waigiensis (Cuvier)
Inhabits rocky or coral reefs, frequently in weedy areas; distinguished by barramundi-like appearance, and glassy eyes, colour ranges from light silvery grey to very dark brown; Fremantle northwards; N. Indian Ocean and W. Pacific; to 47 cm. ★★

343 SPIKY BASS
Hypopterus macropterus (Günther)
Inhabits sand-weed areas; has somewhat similar shape to 342, although a much smaller fish and deeper-bodied; colour is strongly mottled; Jurien Bay to Barrow Island; W. Australia only; to 14 cm.

344 SAILFIN PERCHLET
Ambassis interruptus Bleeker
Inhabits brackish bays and estuaries, also mangrove-lined tidal creeks; distinguished by semi-transparent appearance, tall dorsal fin, and white tips on pelvic fins and front of anal fin; Derby northwards; Indo-Australian Archipelago; to 10 cm.

345 SCALLOPED PERCHLET
Ambassis nalua (Hamilton)
Inhabits brackish bays and estuaries; similar to 344 but has slight hump on snout, dorsal fin is slightly shorter, and lacks white tip on pelvic and anal fins; Admiralty Gulf area; E. Indian Ocean and Indo-Australian Archipelago; to 12 cm.

346 TELKARA PERCHLET
Ambassis vachelli Richardson
Inhabits brackish bays, estuaries and tidal creeks; similar to 344 and 345, but has more slender body and lower dorsal fin; frequently misidentified as *A. dussumieri;* Carnarvon northwards; Indo-Australian Archipelago; to 7 cm.

347 DEEPSEA JEWFISH
Glaucosoma burgeri Richardson
Inhabits trawling grounds; distinguished by silvery appearance, large eye, and single dorsal fin that is elevated on rear portion, frequently with narrow stripes which may disappear in larger fish; closely related to the jewfish *(G. hebraicum)* of southern waters; Onslow northwards; mainly W. Pacific; to 45 cm; 2.5 kg. ★★★

348 THREADFIN PEARL-PERCH
Glaucosoma magnificum (Ogilby)
Inhabits trawling grounds; similar to 347, but has elongate dorsal fin filaments and brown bars on head; Exmouth Gulf northwards; N. Australia and New Guinea; to 32 cm. ★★★

349 YELLOWTAIL TRUMPETER
Amniataba caudovittatus (Richardson)
Inhabits estuaries over sand-weed bottoms; distinguished by yellow fins and conspicuous stripes and spots on tail; also known as Yellowtail grunter and Yellow-tailed perch; Cape Leeuwin northwards; Australia and New Guinea; to 28 cm; .36 kg. ★★

350 TRUMPETER
Pelates quadrilineatus (Bloch)
Inhabits coastal waters, entering estuaries; distinguished by 6-8 straight dark stripes on side and none on tail; Shark Bay northwards; Indo-W. Pacific; to 20 cm. ★

351 CRESCENT PERCH
Terapon jarbua (Forsskål)
Inhabits coastal waters, entering estuaries and lower reaches of freshwater streams; distinguished by 3 or 4 curved dark stripes on side; Shark Bay northwards; Indo-W. Pacific; to 32 cm; .125 kg. ★

352 THREE-LINED GRUNTER
Terapon puta (Cuvier)
Inhabits coastal waters, entering brackish estuaries; distinguished by 3 or 4 straight dark stripes on side; also known as Spiny-cheeked grunter; Exmouth Gulf northwards; N. Indian Ocean and Indo-Australian Archipelago; to 16 cm. ★

353 BANDED GRUNTER
Terapon theraps (Cuvier)
Inhabits coastal waters; similar to 352, but has wider stripes and scales much larger (46-56 in lateral line versus 70-85), also is deeper bodied; Exmouth Gulf northwards; Indo-W. Pacific; to 28 cm. ★

354 GOLDEN-LINED WHITING
Sillago analis Whitley
Inhabits sandy bottoms near shore; distinguished by golden-silver to golden-yellow stripe along middle of side; also known as Rough-scale whiting; Shark Bay northwards; N. Australia and New Guinea; to 45 cm. ★★★

355 STOUT WHITING
Sillago robusta Stead
Inhabits sandy bottoms near shore; distinguished by yellow blotch on cheek and silvery stripe on middle of side; Fremantle northwards, common at Shark Bay, but rare north of there, Australia only; to 30 cm. ★★★

356 TRUMPETER WHITING
Sillago maculata burrus Richardson
Inhabits sandy bottoms near shore; distinguished by irregular dark blotches on side; Geographe Bay northwards; N.W. Australia, New Guinea, and Indonesia; to 30 cm. ★★★

357 NORTHERN WHITING
Sillago sihama (Forsskål)
Inhabits sandy bottoms near shore; a plain uniform coloured whiting without distinguishing marks; also known as Sand smelt; Broome northwards; Indo-W. Pacific; to 31 cm. ★★★

358 WESTERN SCHOOL WHITING
Sillago vittata McKay
Inhabits sandy bottoms near shore; distinguished by dark diagonal lines on back; Geographe Bay to North West Cape; W.A. only; to 30 cm. ★★★

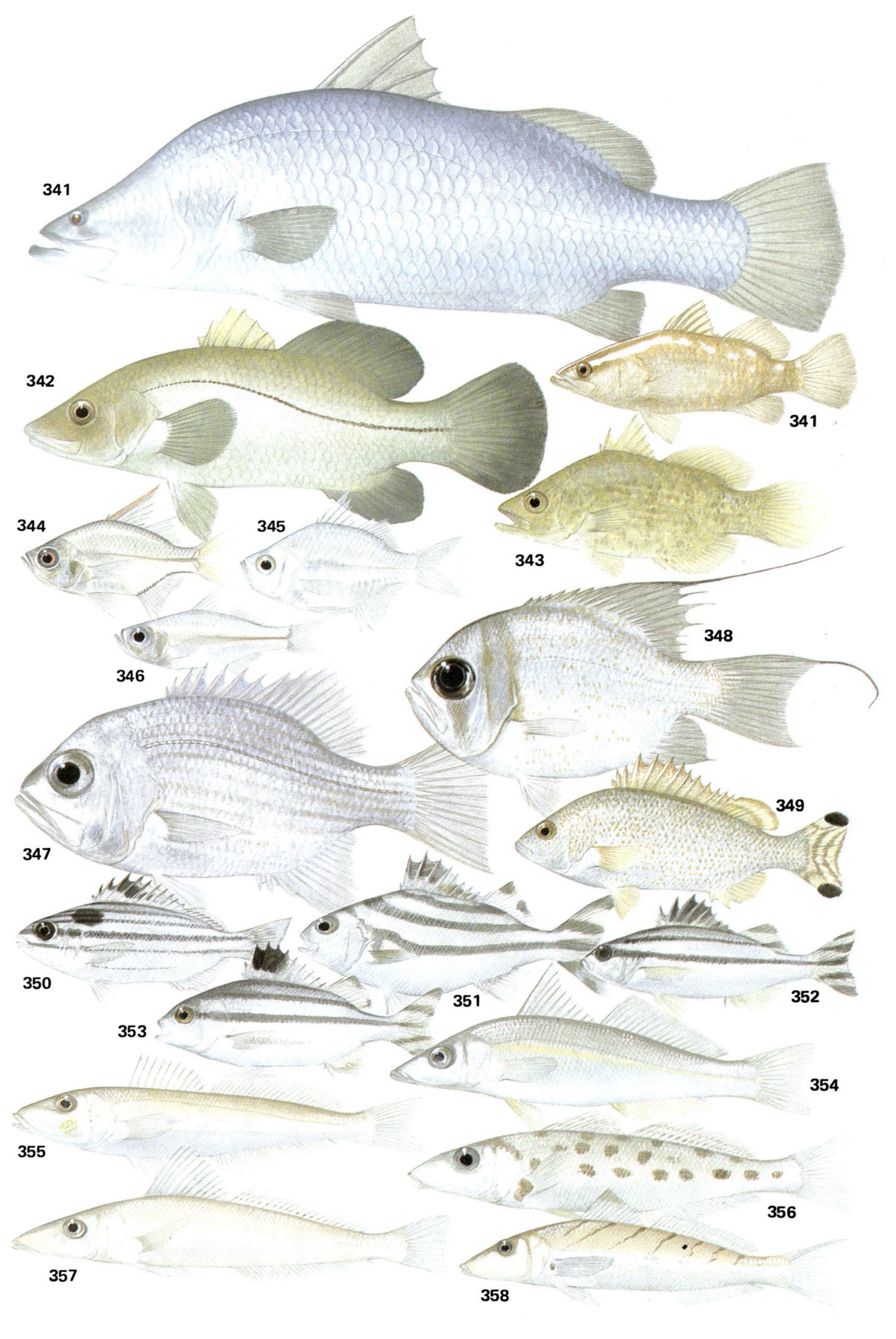

359 WHITE-BARRED BIGEYE
Pristigenys niphonia (Cuvier)
Inhabits deeper offshore reefs and trawling grounds; distinguished by white bars on sides; Barrow Island northwards; Indo-W. Pacific; to 26 cm.   ★★★

360 DUSKYFIN BIGEYE
Priacanthus cruentatus (Lacepède)
Inhabits coral reefs, usually in caves except at night; similar to 362, but has lighter pelvic fin, spots (sometimes faint) on dorsal, anal and tail fins, and tail not crescentic; spot pattern shown is not always evident; also known as Glass bigeye; North West Shelf; Indo-C. Pacific; to 30 cm.   ★★★

361 RED BIGEYE
Priacanthus macracanthus Cuvier
Inhabits inshore and offshore reefs; distinguished by relatively short dorsal and anal fins which are distinctly spotted, similar to 365, which lacks distinct spotting; also known as Large-spined bigeye; Ningaloo Reef northwards; mainly W. edge of Pacific; to 35 cm; 1.021 kg.   ★★★

362 LUNAR-TAILED BIGEYE
Priacanthus hamrur (Forsskål)
Inhabits coral reefs and rocky bottoms; similar to 360, but has darker pelvic fins, crescentic tail, and lacks spotting on fins except a dark spot present at base of pelvic fins; Abrolhos northwards; Indo-C. Pacific; to 40 cm; .4 kg.   ★★★

363 THREADFIN BIGEYE
Priacanthus tayenus Richardson
Inhabits coral reefs and rocky bottoms to at least 200 m depth; distinguished by relatively tall dorsal and anal fins and filamentous tips on tail; Shark Bay northwards; N. Indian Ocean and W. edge of Pacific; to 35 cm.   ★★★

364 ROBUST BIGEYE
Priacanthus sp. 1
Inhabits coral reefs and rocky bottoms; distinguished by tall dorsal and anal fins, large pelvic fins, and dark membrane between first and second dorsal spines; possibly a new species; Derby northwards; Indo-Australian Archipelago; to 40 cm.   ★★★

365 DEEPSEA BIGEYE
Priacanthus sp. 2
Inhabits offshore waters between 150-300 m depth; similar to 361, but lacks distinct spotting on fins; Browse Island northwards; Indo-Australian Archipelago; to 25 cm.   ★★★

366 LONG-FINNED BIGEYE
Cookeolus boops (Bloch & Schneider)
Inhabits trawl grounds and rocky bottoms; distinguished by very large pelvic fins and tall dorsal and anal fins; Shark Bay northwards; Indo-C. Pacific; to 60 cm.   ★★★

367 WOLF CARDINALFISH
Cheilodipterus artus Smith
Inhabits coral reefs, in caves and crevices; distinguished by 8-10 stripes, yellow area with black spot at tail base, and large canine teeth; North West Shelf; Indian Ocean; to 12 cm.

368 EIGHT-LINED CARDINALFISH
Cheilodipterus lineatus Lacepède
Inhabits coral reefs, in caves and crevices; distinguished by striped pattern, lack of yellow on tail base, and large canine teeth; Ningaloo Reef northwards; Indo-C. Pacific; to 22 cm.

369 FIVE-LINED CARDINALFISH
Cheilodipterus quinquelineatus Cuvier
Inhabits coral reefs, in caves and crevices; distinguished by 5 stripes on side, spot at tail base, and large canine teeth; Abrolhos northwards; Indo-C. Pacific; to 12 cm.

370 SAILFIN CARDINALFISH
Pterapogon mirifica (Mees)
Inhabits coral reef caves and ledges; distinguished by blackish colour, white tail and enlarged fins; Ningaloo Reef to Yampi Sound; W. Australia only; to 14 cm.

371 WEED CARDINALFISH
Foa brachygramma (Jenkins)
Inhabits sand-weed areas, usually under dead coral slabs; distinguished by mottled pattern, incomplete lateral line and presence of teeth on palate; Ningaloo Reef northwards; Indo-C. Pacific; to 6 cm.

372 VARIEGATED CARDINALFISH
Fowleria variegatus (Valenciennes)
Inhabits coral reef crevices; distinguished by spot on gill cover, and irregular spotting on body and fins; Abrolhos northwards; Indo-C. Pacific; to 8 cm.

373 AURITA CARDINALFISH
Fowleria aurita (Valenciennes)
Inhabits coral reef crevices; distinguished by spot on gill cover and lack of markings on body and fins; Abrolhos northwards; Indo-C. Pacific; to 8 cm.

374 STRIPED SIPHONFISH
Siphamia majimai Matsubara & Iwai
Inhabits coral reefs, usually found among the spines of sea urchins; distinguished by silver stripe on belly and alternating black and white stripes on side or may be entirely blackish; Exmouth Gulf northwards; mainly W. Pacific; to 5 cm.

375 PINK-BREASTED SIPHONFISH
Siphamia roseigaster (Ramsay & Ogilby)
Inhabits coastal reefs; distinguished by silver stripe on belly, plain pale colouration and rosy fins; Port Hedland northwards; N. Australia only; to 7 cm.

BIG EYES

The larger fishes (359-366) on this plate are members of the family Priacanthidae, also known as bigeyes. They are somewhat similar in appearance to the squirrelfishes and soldierfishes (Plate 12), but are distinguished by smaller scales, a larger, upturned mouth, and lack of spines on the head. They are also similar to squirrelfishes in behaviour, being essentially nocturnal and spending the daylight hours in caves. At night they emerge to feed on cephalopods, crustaceans, and fish. Several species occur in shallow reef habitats while others range to depths as great as 500 m. Colour patterns are generally bright red, but in some species the livery can be changed instantaneously to a dull silver-grey. The approximately 20 known species are mainly found in the Indo-Pacific region.

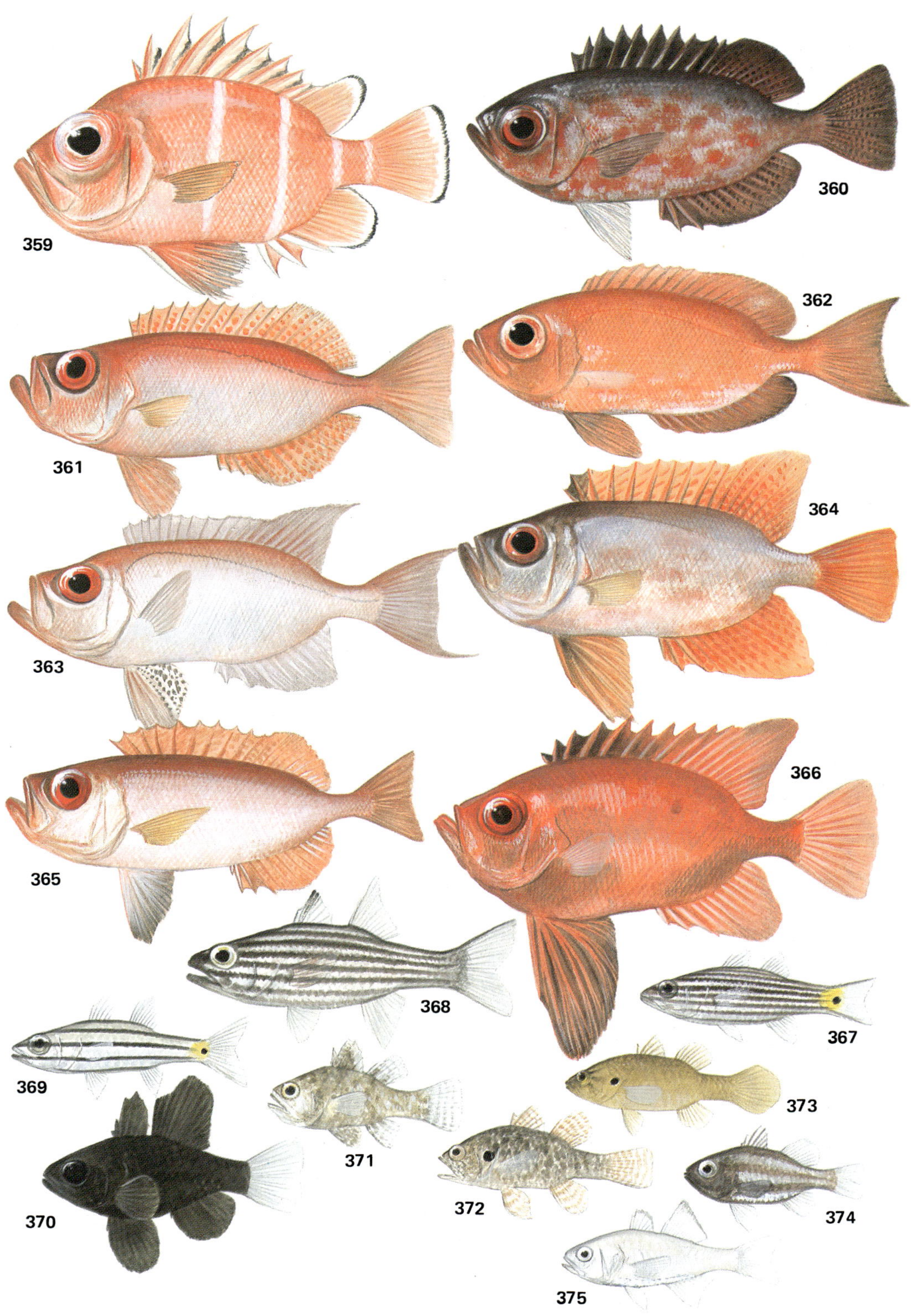

PLATE 24: CARDINALFISHES

376 STRIPED CARDINALFISH
Apogon angustata Smith & Radcliffe
Inhabits coral reef crevices; distinguished by relatively thin black stripes and distinct spot at base of tail; similar to 377, but found in deeper water; Ningaloo Reef northwards; mainly Indo-C. Pacific; to 9 cm.

377 REEF-FLAT CARDINALFISH
Apogon taeniophorus Regan
Inhabits shallow reef flats exposed to wave action; similar to 376, but no black spot at tail base; Shark Bay northwards; W. Pacific and E. Indian Ocean; to 10 cm.

378 DOEDERLEIN'S CARDINALFISH
Apogon doederleini Jordan & Snyder
Inhabits coral reef crevices and southern rocky reefs; distinguished by thin red to black stripes on side and spot at tail base; Abrolhos northwards; mainly W. Pacific; to 9 cm.

379 CANDYSTRIPE CARDINALFISH
Apogon endekataenia Bleeker
Inhabits coral reef crevices; similar to 380, but stripes are thinner and lacks short (incomplete) stripe behind upper corner of eye; Ningaloo Reef northwards; Indo-C. Pacific; to 10 cm.

380 COOK'S CARDINALFISH
Apogon cooki Mackay
Inhabits shallow reefs; frequently in rocky crevices in weedy areas; distinguished from other striped species by incomplete stripe behind upper corner of eye; Jurien Bay northwards; Indo-W. Pacific; to 10 cm.

381 BLACK-TIPPED CARDINALFISH
Apogon semilineatus Temminck & Schlegel
Inhabits offshore trawling grounds; distinguished by black tip on dorsal fin and pair of narrow stripes along back; Broome northwards; mainly W. Pacific; to 10 cm.

382 PALE-STRIPED CARDINALFISH
Apogon pallidofasciatus Allen
Inhabits shallow inshore reefs, in crevices often located in weedy areas; distinguished by dusky brown colour with faint stripes on side; North West Cape northwards; N. Australia only; to 13 cm.

383 BROAD-BANDED CARDINALFISH
Apogon fasciatus (Shaw)
Inhabits flat sand bottoms with rocky outcrops; distinguished by dark mid-lateral stripe and 1-2 thinner stripes above; Shark Bay northwards; Indo-W. Pacific; to 10 cm.

384 EVERMANN'S CARDINALFISH
Apogon evermanni Jordan & Snyder
Inhabits deeper offshore reefs; distinguished by red colouration, dark stripe behind eye, and small white spot behind second dorsal fin; Port Hedland northwards; worldwide circumtropical; to 9 cm.

385 THREE-SADDLE CARDINALFISH
Apogon bandanensis Bleeker
Inhabits coral reefs, frequently among branching corals; similar to 386, but dark bar completely encircles tail base; Ningaloo Reef northwards; mainly W. & C. Pacific; to 10 cm.

386 SAMOAN CARDINALFISH
Apogon savayensis Günther
Inhabits coral reefs; similar to 385, but bar at tail base not complete; Ningaloo Reef northwards; Indo-C. Pacific; to 10 cm.

387 SEVEN-BANDED CARDINALFISH
Apogon septemstriatus Günther
Inhabits flat sand bottoms and offshore trawling grounds; similar to 383, but has stripe on midline of forehead from snout to dorsal fin; Dampier Archipelago northwards; W. Pacific and E. Indian Ocean; to 9 cm.

388 SPINY-EYED CARDINALFISH
Apogon fraenatus Valenciennes
Inhabits coral reef crevices; distinguished by black stripe along middle of side and spot at tail base; Ningaloo Reef northwards; Indo-C. Pacific; to 10 cm.

389 BLUE-STRIPED CARDINALFISH
Apogon cyanosoma Bleeker
Inhabits shallow reef crevices; distinguished by yellow-orange stripes; Jurien Bay northwards; Indo-W. Pacific; to 8 cm.

390 IRIDESCENT CARDINALFISH
Apogon kallopterus Bleeker
Inhabits caves and coral reef crevices; distinguished by broad dark stripe along middle of side, dusky colour on back, and spot at base of tail; North West Cape northwards; Indo-C. Pacific; to 15 cm.

391 MANY-LINED CARDINALFISH
Apogon chrysotaenia Bleeker
Inhabits coral reef crevices; distinguished by blue stripes on head, faint pale stripes on side, and spot on tail base; Ningaloo Reef northwards; Indo-Australian Archipelago; to 10 cm.

392 GOBBLEGUTS
Apogon ruppelli Günther
Inhabits inshore reefs and weedy areas, also in estuaries; distinguished by row of small black spots on upper side; Albany northwards; Australia and New Guinea; to 12 cm; 0.030 kg.

393 TIMOR CARDINALFISH
Apogon timorensis Bleeker
Inhabits inshore reef flats with weed and sand, found under rocks; distinguished by overall dusky colour, sometimes with faint dark bars, and a narrow stripe below eye; Ningaloo Reef northwards; Indo-W. Pacific; to 8 cm.

CARDINALFISHES

Cardinalfishes (family Apogonidae) are represented on Plates 23-25, nos. 367-411). They are small, reef dwelling fishes that inhabit tropical and temperate seas. Some species also frequent estuaries and freshwater streams. The majority of the estimated 200 species are found on Indo-Pacific coral reefs. Most are nocturnally active and spend daylight hours in the dark recesses of caves and ledges. They mainly feed on fishes and crustaceans. A few species, for example members of the genus *Rhabdamia* (408-409), form dense, open-water shoals, that literally envelope entire coral formations. Male cardinalfishes incubate a fertilised egg mass in their mouth for several days until hatching.

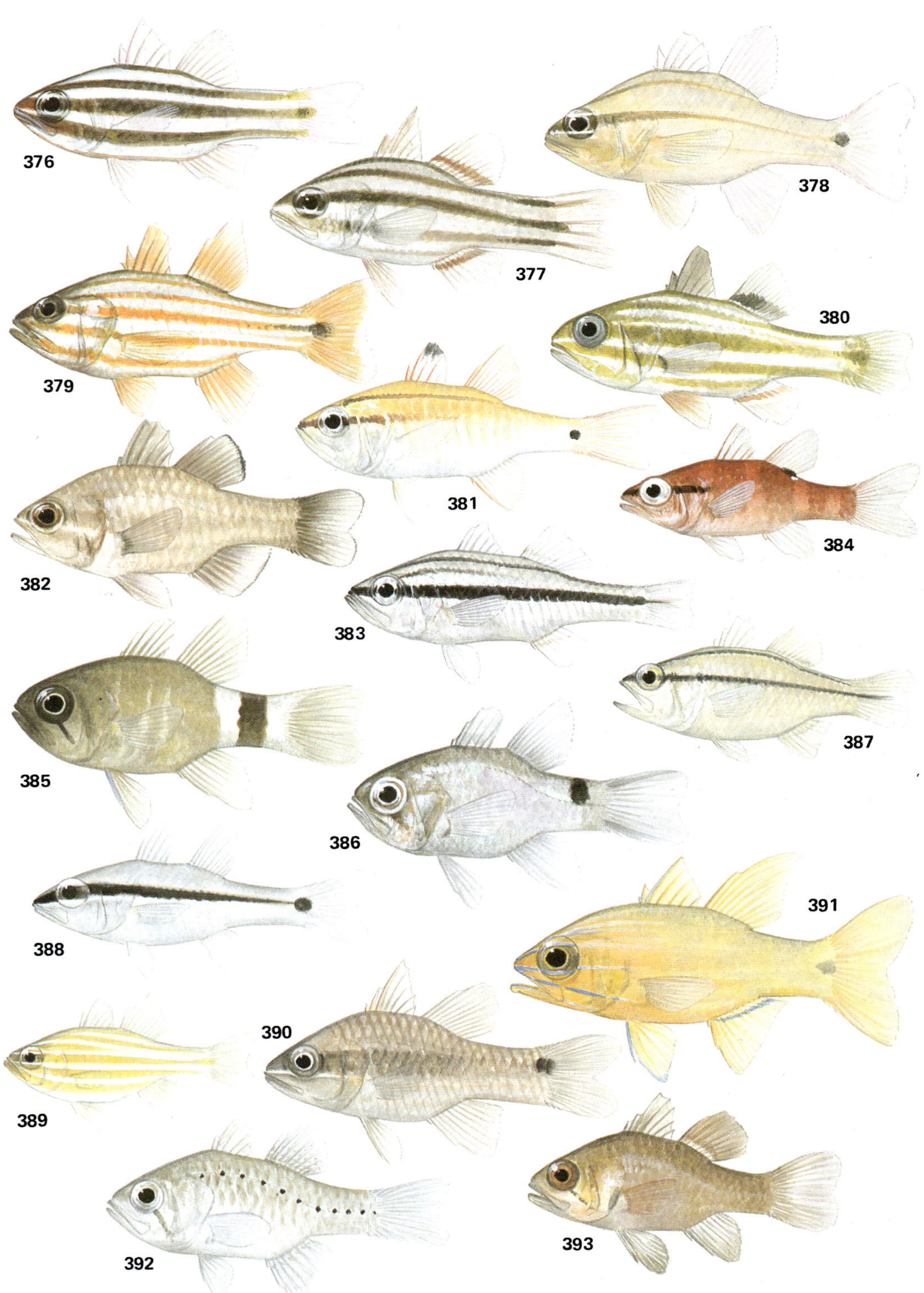

376
377
378
379
380
381
382
383
384
385
386
387
388
389
390
391
392
393

394 MOLUCCAN CARDINALFISH
Apogon moluccensis Valenciennes
Inhabits coral reef crevices; distinguished by broad, faint stripe on side and pearly spot just behind dorsal fin (not shown), Ningaloo Reef northwards; mainly Indo-Australian Archipelago; to 9 cm.

395 YELLOW CARDINALFISH
Apogon virgalatus Allen and Randall
Inhabits silty or sandy areas, around rocky outcrops; distinguished by yellow-orange stripes on side and black spot on tail base; the species has not yet been scientifically described; Dampier Archipelago northwards; Indo-Australian Archipelago; to 8 cm.

396 OBLIQUE-BANDED CARDINALFISH
Apogon semiornatus Peters
Inhabits coral reef caves and crevices; distinguished by semi-transparent appearance and oblique dark bands; Abrolhos northwards; Indo-C. Pacific; to 7 cm.

397 TWO-EYED CARDINALFISH
Apogon nigripinnis Cuvier
Inhabits trawling grounds and inshore reefs; distinguished by dark colour and pale-edged black spot on middle of side; Shark Bay northwards; Indo-W. Pacific; to 8 cm.

398 LITTLE RED CARDINALFISH
Apogon coccineus Rüppell
Inhabits coral reef crevices; distinguished by semi-transparent appearance and overall red colour, similar to 399, but tail base is shorter; Abrolhos northwards; Indo-C. Pacific; to 5 cm.

399 BIG RED CARDINALFISH
Apogon crassiceps Garman
Inhabits coral reef crevices; similar to 398, but grows much larger and has longer tail base; Dampier Archipelago northwards; Indo-C. Pacific; to 12 cm.

400 MANY-BANDED CARDINALFISH
Apogon brevicaudatus Weber
Inhabits trawling grounds and inshore reefs; distinguished by 8-9 narrow stripes on side and black spot at base of second dorsal fin; Shark Bay northwards; N. Australia and New Guinea; to 9 cm.

401 THREE-SPOT CARDINALFISH
Apogon trimaculatus Cuvier
Inhabits coral reef crevices; distinguished by pair of faint bars, one below each dorsal fin, and dark spot on tail base; Ningaloo Reef northwards; mainly W. Pacific; to 15 cm.

402 FLAGFIN CARDINALFISH
Apogon ellioti Day
Inhabits trawling grounds; distinguished by black outer half of first dorsal fin and stripe across middle of anal and second dorsal fins; Broome northwards; Indo-W. Pacific; to 12 cm.

403 PEARLY-FINNED CARDINALFISH
Apogon poecilopterus Cuvier
Inhabits trawling grounds; similar to 404, but lacks dark spot at rear base of second dorsal fin; Exmouth Gulf northwards; mainly W. Pacific; to 12 cm.

404 OCELLATED CARDINALFISH
Apogon carinatus Cuvier
Inhabits trawling grounds; distinguished by dark spot at rear base of second dorsal fin; Broome northwards; mainly W. Pacific, to 12 cm.

405 RING-TAILED CARDINALFISH
Apogon aureus (Lacepède)
Inhabits coral reef caves and crevices; distinguished by golden-orange colour and black ring around tail base; Abrolhos northwards; Indo-C. Pacific; to 15 cm; 0.087 kg.

406 SPINDLE-EGG CARDINALFISH
Apogon melanopus (Weber)
Inhabits trawling grounds and inshore reefs; distinguished by dark bar below dorsal fin and one or more rows of spots on rear half of body, these markings are often faint; this fish has a very unusual spindle-shaped egg; Derby northwards; N. Australia and Arafura Sea; to 12 cm; *Apogon fusovatus* Allen is a synonym.

407 CREAM-SPOTTED CARDINALFISH
Apogon albimaculosus Kailola
Inhabits trawling grounds; distinguished by cluster of large white spots on side and dark spotting on fins; Broome northwards; N. Australia and New Guinea; to 10 cm.

408 BLACK-NOSED CARDINALFISH
Rhabdamia cypselurus Weber
Inhabits coral reefs, forming large aggregations; distinguished by transparent appearance, similar to 409 but has black stripe on side of snout and dark margins on tail; Ningaloo Reef northwards; Indo-C. Pacific; to 6 cm.

409 SLENDER CARDINALFISH
Rhabdamia gracilis Bleeker
Inhabits coral reefs, forming large aggregations; similar to 408, but lacks stripe on snout and dark margins on tail (sometimes has black spot at tips of lobes); Ningaloo Reef northwards; Indo-C. Pacific; to 6 cm.

410 NARROW-LINED CARDINALFISH
Archamia fucata (Cantor)
Inhabits coral reefs and inshore rocky areas; similar to 411, but has spot on tail base and lacks dark blotch above upper edge of gill cover; Ningaloo Reef northwards; Indo-C. Pacific; to 9 cm.

411 BLACKSPOT CARDINALFISH
Archamia melasma Lachner & Taylor
Inhabits coral reefs and inshore rocky areas; similar to 410, but has dark blotch above upper edge of gill cover and lacks spot on tail base; Dampier Archipelago northwards; N. Australia and New Guinea; to 9 cm.

NEW DISCOVERIES

One of the exciting aspects of the Australian fish fauna is that many new species have been discovered in recent years, including a significant number from northwestern waters. More efficient methods of collecting including the widespread application of SCUBA techniques and increased ease of access to remote areas are contributing factors in these discoveries. Several of the cardinalfishes on Plates 24 and 25 were discovered and subsequently described for the first time during the past 10-12 years including 382, 406, and 407. An additional species (395) was only collected recently and is still awaiting description.

412 PENNANTFISH
Alectes ciliaris (Bloch)
Inhabits coastal reefs; juvenile has long filamentous fin rays, both juvenile and adult distinguished from 413 by more rounded head profile and eye closer to mouth; Geographe Bay northwards; worldwide tropical seas; to 130 cm; 13 kg. ★★

413 DIAMOND TREVALLY
Alectes indicus (Rüppell)
Inhabits coastal reefs; similar to 412, but has more angular head profile and wider space between eye and mouth, both species differ from other trevallies by their scaleless skin; Shark Bay northwards; Indo-W. Pacific; to 150 cm. ★★

414 FRINGE-FINNED TREVALLY
Absalom radiatus (Macleay)
Inhabits coastal waters, sometimes in estuaries or river mouths; distinguished by yellow tail with black upper tip; female is shown here, male (Plate 27, 445) has long filamentous rays on the dorsal and anal fins; Port Hedland northwards; Indo-Australian Archipelago; to 40 cm. ★★

415 SMALL MOUTH SCAD
Alepes sp.
Inhabits coastal waters; distinguished by clear fleshy eyelid covering rear half of eye and often has dark tips on lobes of tail; Exmouth Gulf northwards; Indo-Australian Archipelago; to 35 cm. ★★

416 YELLOWTAIL SCAD
Atule mate (Cuvier)
Inhabits coastal waters, forming large schools; distinguished by clear fleshy eyelid covering most of eye except narrow slit in centre; Exmouth Gulf northwards; Indo-C. Pacific; to 30 cm; .087 kg. ★★

417 CLUB-NOSED TREVALLY
Carangoides chrysophrys (Cuvier)
Inhabits coastal waters; distinguished by gently sloping head profile except abruptly vertical at tip of snout and has scaleless area on breast extending on to pectoral fin base; Exmouth Gulf northwards; Indo-W. Pacific; to 44 cm. ★★

418 ONION TREVALLY
Carangoides caeruleopinnatus (Rüppell)
Inhabits coastal waters; distinguished by relatively deep body; short dorsal and anal fin lobes, small black blotch on upper margin of gill cover, and small yellow spots on body; Exmouth Gulf northwards; Indo-C. Pacific; to 40 cm. ★★

419 BLUE TREVALLY
Carangoides ferdau (Forsskål)
Inhabits coastal waters and offshore reefs; distinguished by separated scaleless areas on breast and base of pectoral fin, a bluntly rounded snout, and frequently has 5-6 dusky bars on sides; Ningaloo Reef northwards; Indo-C. Pacific; to 70 cm. ★★

420 GOLD-SPOTTED TREVALLY
Carangoides fulvoguttatus (Forsskål)
Inhabits coastal waters; distinguished by relatively elongate shape and many gold or brassy spots on side (mainly on back), similar to 426 but has eye higher above mouth, more tapered snout, and more yellow spots; also known as Turrum and Yellow-spotted trevally; Cape Leeuwin northwards; Indo-W. Pacific; to 130 cm; 11.6 kg. ★★★

421 WHITEFIN TREVALLY
Carangoides equula (Temminck & Schlegel)
Inhabits coastal waters; distinguished by blackish or dusky submarginal band on second dorsal fin and sometimes on anal fin, by very short lobes at front of dorsal and anal fins, and a fully scaled breast; Shark Bay northwards; Indo-W. Pacific; to 37 cm. ★★

422 EPAULET TREVALLY
Carangoides humerosus (McCulloch)
Inhabits coastal waters; distinguished by large eye (about equal to distance from eye to snout tip); blackish first dorsal fin, and moderately long dorsal and anal lobes; Derby northwards; Indo-Australian Archipelago; to 25 cm. ★★

423 BUMP-NOSED TREVALLY
Carangoides hedlandensis (Whitley)
Inhabits coastal waters; distinguished by long filamentous extensions on dorsal and anal fin rays, this feature also present in adult males of 414, but that species much more slender and has black tip on upper lobe of tail; also known as Port Hedland trevally; Exmouth Gulf northwards; Indo-C. Pacific; to 32 cm; .25 kg. ★★

424 WHITE-TONGUED TREVALLY
Carangoides talamparoides Bleeker
Inhabits coastal waters; distinguished by a white or pale grey tongue, relatively deep body, steep head profile, and extensive scaleless area encompassing breast, pectoral fin base, and small area above pectoral fin; Dampier northwards; N. Indian Ocean and Indo-Australian Archipelago; to 32 cm. ★★

425 THICKLIP TREVALLY
Carangoides orthogrammus Jordan & Gilbert
Inhabits coastal areas; distinguished by several ovate yellow spots on middle of side and well separated scaleless areas on breast and base of pectoral fin; also known as False bluefin trevally; Fremantle northwards; Indo-C. Pacific; to 70 cm; 3.5 kg. ★★

426 BLUDGER TREVALLY
Carangoides gymnostethus (Cuvier)
Inhabits coastal areas in the vicinity of coral or rocky reefs; distinguished by relatively elongate body and a few brown or golden spots often present on side, similar to 420, but has eye closer to level of mouth, fewer yellow spots, and a steeper snout profile; Coral Bay northwards; Indo-W. Pacific; to 90 cm; 11.113 kg. ★★

427 MALABAR TREVALLY
Carangoides malabaricus (Bloch & Schneider)
Inhabits coastal waters; similar to 424, but tongue grey-brown to brown instead of whitish; Exmouth Gulf northwards; Indo-W. Pacific; to 28 cm. ★★

428 JAPANESE TREVALLY
Carangoides uii Wakiya
Inhabits coastal waters; distinguished by relatively deep body and thread-like filament at front of second dorsal (and often anal) fin; Broome northwards; Indo-W. Pacific to 25 cm. ★★

PLATE 27

429 GIANT TREVALLY
Caranx ignobilis (Forsskål)
Inhabits coastal and offshore waters in the vicinity of reefs; the largest of the trevallies, distinguished by steep forehead profile and silvery to dusky colouration; also known as Lowly trevally; Rottnest Island northwards; Indo-C. Pacific; to 170 cm; 34.6 kg. ★★

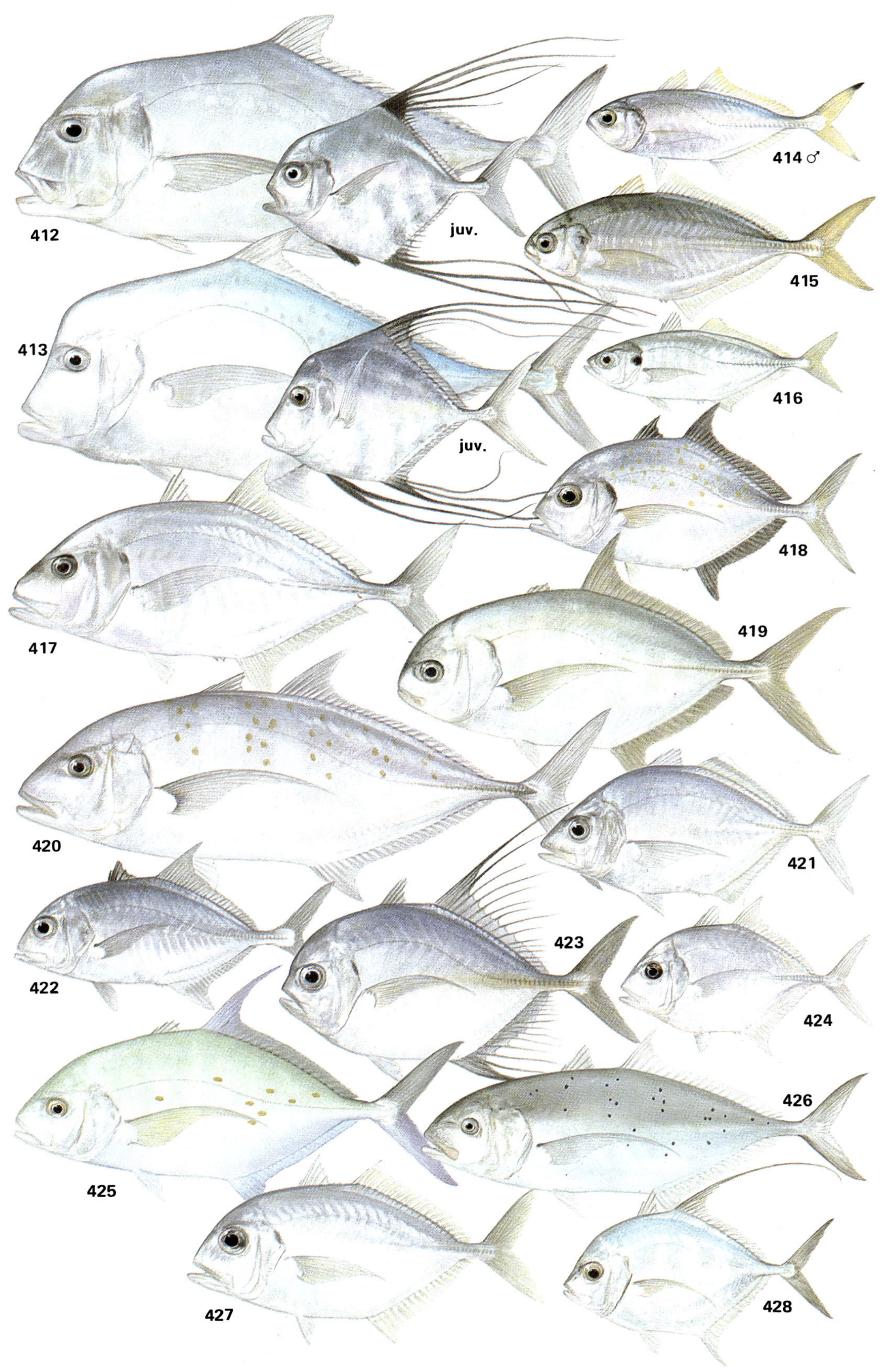

412
juv.
414 ♂
415
413
416
juv.
418
417
419
420
421
422
423
424
425
426
427
428

429 GIANT TREVALLEY
Caranx ignobilis (Forsskål). Text on page 70.

430 BLACK TREVALLY
Caranx lugubris Poey
Inhabits mainly offshore waters in the vicinity of coral reefs; distinguished by dark colouration ranging from brown to nearly black; Dampier Archipelago northwards; worldwide tropical seas; to 80 cm. ★★

431 BLUEFIN TREVALLY
Caranx melampygus Cuvier
Inhabits coastal and offshore waters in the vicinity of reefs; distinguished by blue fins and dark speckling on upper half of body; Ningaloo Reef northwards; Indo-E. Pacific; to 100 cm; 1.56 kg. ★★★

432 BRASSY TREVALLY
Caranx papuensis Alleyne & Macleay
Inhabits coastal and offshore waters in the vicinity of reefs; similar to 429, but forehead not as steep and has white margin on lower lobe of tail, also usually scattered dark spots on upper side; also known as Papuan trevally; Exmouth Gulf northwards; E. Indian Ocean and W. Pacific; to 75 cm; 2.5 kg. ★★

433 TILLE TREVALLY
Caranx tille Cuvier
Inhabits coastal and offshore waters in the vicinity of reefs; distinguished by relatively slender shape, well developed gelatinous membrane covering much of eye, and blackish spot on upper corner of gill cover; Port Hedland northwards; Indo-W. Pacific; to 70 cm. ★★

434 BANDED SCAD
Caranx para Cuvier
Inhabits coastal waters; distinguished by small size, exaggerated profile of belly (i.e. ventral profile more convex than dorsal profile) and black spot on upper corner of gill cover; Port Hedland northwards; Indo-W. Pacific; to 18 cm. ★★

435 BLUE-SPOTTED TREVALLY
Caranx bucculentus Alleyne & Macleay
Inhabits coastal waters; distinguished by steep forehead profile, large eye, dark spot at upper pectoral fin base, and blue spots on upper side, juveniles with 6 dark bars which develop into 3 horizontal rows of square blotches; Exmouth Gulf northwards; N. Australia and New Guinea; to 66 cm; 2.948 kg. ★★

436 BIGEYE TREVALLY
Caranx sexfasciatus Quoy & Gaimard
Inhabits coastal and offshore waters in the vicinity of reefs; distinguished by relatively large eye with well developed gelatinous membrane and white tip on dorsal fin lobe, juveniles have 5-6 dark bars on body and sometimes occur in fresh water and estuaries; Geographe Bay northwards; Indo-E. Pacific; to 78 cm; 4.01 kg. ★★

437 ROUGHEAR SCAD
Decapterus tabl Berry
Inhabits coastal waters, occurring in schools; a slender, silvery fish distinguished from other mackerel scad by a red tail and 4-10 scales in straight part of lateral line immediately preceding expanded bony scutes on rear part of body; Abrolhos northwards; Atlantic and Indo-C. Pacific; to 50 cm. ★★

438 REDTAIL SCAD
Decapterus kurroides Bleeker
Inhabits coastal waters, occurring in schools; similar to 437, including red tail, but lacks a straight section in the lateral line (i.e. between curved anterior part and expanded bony scules on rear portion); Port Hedland northwards; Indo-W. Pacific; to 50 cm. ★★

439 MACKEREL SCAD
Decapterus macarellus Cuvier
Inhabits coastal waters, occurring in schools; distinguished by yellow-green tail and 18-32 scales in straight part of lateral line in front of bony scutes; Broome northwards; worldwide tropical seas; to 32 cm; 1.65 kg. ★★

440 RUSSELL'S MACKEREL SCAD
Decapterus russelli (Rüppell)
Inhabits coastal waters, occurring in schools; distinguished by clear to dusky tail and 0-4 scales in straight part of lateral line in front of bony scutes; also known as Indian scad; Shark Bay northwards; Indo-W. Pacific; to 38 cm; 1.8 kg. ★★★

441 LONG-BODIED SCAD
Decapterus macrosoma (Bleeker)
Inhabits coastal waters, occurring in schools; distinguished by clear to dusky tail and 14-29 scales in straight part of lateral line in front of bony scutes; Exmouth Gulf northwards; Indo-E. Pacific; to 32 cm. ★★

442 GOLDEN TREVALLY
Gnathanodon speciosus (Forsskål)
Inhabits coastal and offshore waters, usually near reefs; sometimes occurring in schools; distinguished by large fleshy lips, lack of discernible teeth (unlike other trevallies), and golden belly, juvenile (not shown) is bright yellow with dark bars and dark tips on tail; Denmark northwards, but generally rare south of Shark Bay, Indo-E. Pacific; to 111 cm; 15 kg. ★★★

443 FINNY SCAD
Megalaspis cordyla (Linnaeus)
Inhabits coastal waters; distinguished by series of separate finlets behind dorsal and anal fins and greatly expanded bony scutes along middle of side; Fremantle northwards; Indo-C. Pacific; to 80 cm; 1.6 kg. ★★

444 PILOT FISH
Naucrates ductor Linnaeus
Inhabits oceanic waters, usually in company with sharks, rays, turtles, or large fishes; juveniles may occur in floating weed or with jellyfishes; distinguished by prominent dark bars; entire coast of W.A.; worldwide tropical seas; to 70 cm; .72 kg. ★★

445 FRINGE-FINNED TREVALLY
Absalom radiosus (Macleay)
Inhabits coastal waters; distinguished by orange-yellow dorsal and tail fins, and black tip on upper lobe of tail, male (shown here) has filamentous dorsal and anal rays female shown on Plate 26 (414); Port Hedland northwards; Australia only; to 38 cm. ★★

446 BLACK POMFRET
Parastromateus niger (Bloch)
Inhabits offshore waters of continental shelf, frequently in schools; distinguished by equal-shaped dorsal and anal fin with triangular anterior lobe, colour ranges from silvery-grey to bluish-brown; Derby northwards; Indo-W. Pacific; to 55 cm 2.55 kg.

447 SILVER TREVALLY
Pseudocaranx dentex. Text on page 74.

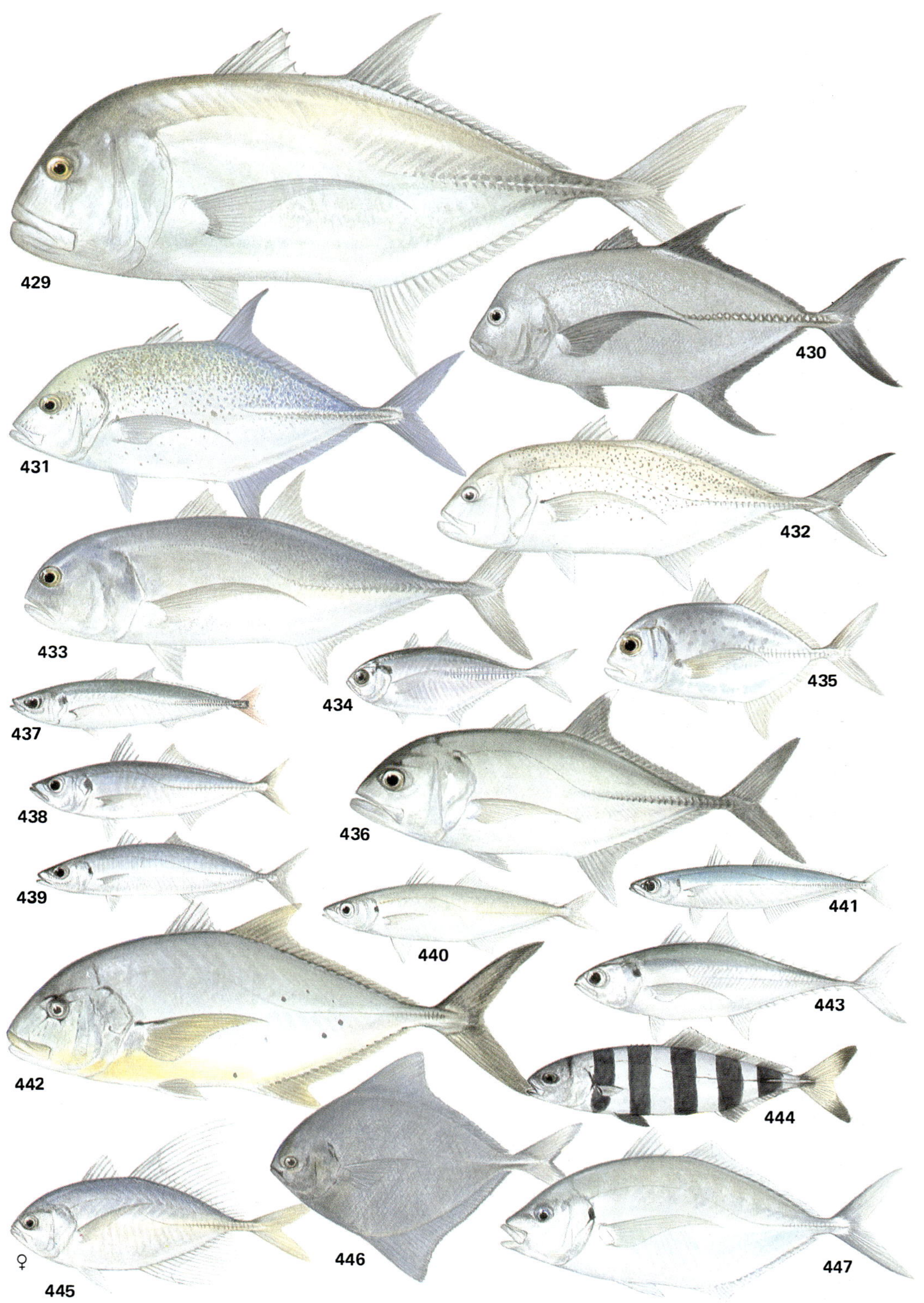

429
430
431
432
433
434
435
436
437
438
439
440
441
442
443
444
445
446
447
♀

448 COMMON DOLPHINFISH
Coryphaena hippurus (Linnaeus)
Inhabits mainly oceanic waters well offshore; distinguished by elongate, compressed body, long-based dorsal and anal fins, and steep forehead profile; female (not shown) is less colourful and lacks a distinct hump on the forehead; also known as Mahimahi and Dorado; almost entire coast of W.A.; worldwide tropical and subtropical seas; to 200 cm; 22.4 kg. ★★★

449 RAINBOW RUNNER
Elegatis bipunnulata (Quoy & Gaimard)
Inhabits coastal waters and offshore reefs, usually in schools; distinguished by slender shape, yellow and blue stripes, and isolated finlets on tail base, Rottnest Island northwards; worldwide tropical and subtropical seas; to 120 cm; 4.76 kg. ★★★

450 NEEDLESKIN QUEENFISH
Scomberoides tol (Cuvier)
Inhabits coastal waters, often in small schools; distinguished by 5-8 round spots, the first 4-5 making contact with lateral line; also known as Slender leatherskin; Cape Cuvier northwards; Indo-W. Pacific; to 60 cm. ★★

451 BARRED QUEENFISH
Scomberoides tala (Cuvier)
Inhabits coastal waters; distinguished by 4-8 vertically elongate blotches most of which make contact with lateral line; also known as Deep leatherskin; Dampier northwards; Indian Ocean and W. edge of Pacific; to 75 cm. ★★

452 TALANG QUEENFISH
Scomberoides commersonnianus Lacepède
Inhabits coastal waters, usually near reefs, but occasionally in estuaries; distinguished by 5-8 blotches that are mainly above lateral line (first two may contact lateral line); also known as Leatherskin; Exmouth Gulf northwards; Indo-W. Pacific; to 120 cm; 11.4 kg. ★★

453 DOUBLE-SPOTTED QUEENFISH
Scomberoides lysan (Forsskål)
Inhabits coastal waters, often in schools; distinguished by double row of 6-8 dark spots on side; Shark Bay northwards; Indo-C. Pacific; to 70 cm. ★★

454 OXEYE SCAD
Selar boops (Valenciennes)
Inhabits coastal waters in large schools; distinguished by large eye covered over by clear fleshy eyelid except for central portion, and by yellow or bronze hue on back; Dampier Archipelago northwards; W. Pacific and E. Indian Ocean; to 25 cm. ★★

455 PURSE-EYED SCAD
Selar crumenthalmops (Bloch)
Inhabits coastal waters in large schools; similar to 454, but lacks bronze or yellow hue, instead usually has broad yellowish mid-lateral stripe; Onslow northwards; worldwide tropical seas; to 30 cm.

456 SMOOTH-TAILED TREVALLY
Selaroides leptolepis Valenciennes
Inhabits coastal waters forming large schools over soft sand or mud bottoms; similar to 454 and 455, but has smaller plate-like scales (scutes) along rear part of lateral line, lacks teeth in upper jaw, and has clear fleshy eyelid covering rear half of eye only; Shark Bay northwards; N. Indian Ocean and W. edge of Pacific; to 20 cm; .06 kg. ★★

457 BLACK-SPOTTED DART
Trachinotus bailloni (Lacepède)
Inhabits coastal waters, frequently in surge off sandy beaches; distinguished by strongly forked tail and 1-5 small black spots along middle of sides; Lancelin northwards; Indo-C. Pacific; to 54 cm; .775 kg. ★★

458 SNUB-NOSED DART
Trachinotus blochii (Lacepède)
Inhabits coastal waters, frequently in surge off sandy beaches; distinguished by strongly forked tail, broadly rounded snout profile, and lack of spots on side; Ningaloo Reef northwards; Indo-C. Pacific; to 65 cm; 9.360 kg. ★★

459 COMMON DART
Trachinotus botla (Shaw)
Inhabits coastal waters, frequently in surge off sandy beaches; distinguished by strongly forked tail and 1-5 large spots along middle of side; Bunbury northwards; Indo-W. Pacific; to 61 cm; 3.504 kg. ★★

460 BLACK-BANDED KINGFISH
Seriolina nigrofasciata (Ruppell)
Inhabits oceanic waters usually well offshore; distinguished by forward slanting dark bars or large blotches arranged in ventral rows; Shark Bay northwards; Indo-W. Pacific; to 70 cm. ★★

461 ALMACO JACK
Seriola rivoliana (Valenciennes)
Inhabits oceanic waters usually well offshore, occasionally visiting coastal areas; distinguished by diagonal dark band above eye and has deeper body than other *Seriola;* North West Cape northwards; worldwide tropical and temperate seas; to 70 cm; .815 kg. ★★

462 YELLOWTAIL KINGFISH
Seriola lalandi Valenciennes
Inhabits coastal and offshore waters, sometimes in schools; similar to 463, but has narrower upper jaw and yellow tail; Shark Bay northwards; worldwide temperate and tropical seas; to 180 cm; 47.250 kg. ★★

463 AMBERJACK
Seriola dumerili (Risso)
Inhabits mainly offshore waters in the vicinity of reefs, sometimes adjacent to dropoffs; similar to 462, but lacks yellow tail and rear part of jaw broadly expanded; Albany northwards; Atlantic and Indo-C. Pacific; to 188 cm; 39.46 kg. ★★

PLATE 27

447 SILVER TREVALLY
Pseudocaranx dentex (Bloch & Schneider)
Inhabits coastal waters, including estuaries, usually in schools; silvery white to pale bluish with or without diffuse dark bars on side, back greenish or bronzy, a dark spot on upper corner of gill cover, and sometimes a yellow stripe on middle of sides; common in southern waters, but rare north of Shark Bay, ranges to North West Cape; Atlantic and Indo-C. Pacific; to 94 cm; 10 kg. ★★

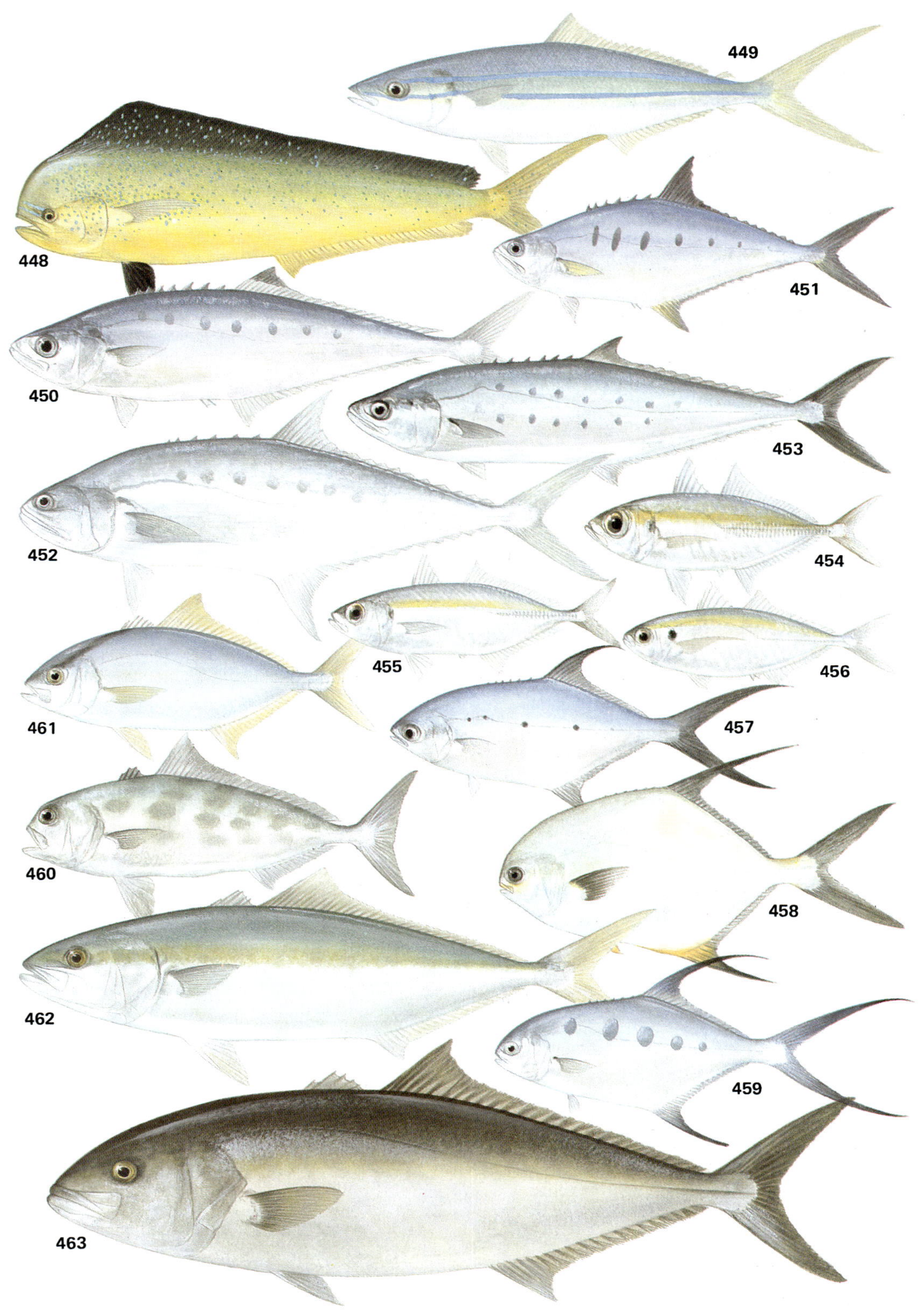

448
449
450
451
452
453
454
455
456
457
458
459
460
461
462
463

464 BLACK-CRESTED TREVALLY
Ulua aurochs Ogilby
Inhabits coastal waters; distinguished by extremely long gill rakers that project into mouth along side of tongue and prominent chin which protrudes well ahead of upper jaw in older specimens (not shown), similar to 465, but has longer dorsal fin filament, diffuse dark bars on side and deeper body; Broome northwards; N. Australia and New Guinea; to 50 cm. ★★

465 CALE CALE TREVALLY
Ulua mentalis (Cuvier)
Inhabits coastal waters; similar to 464, but has shorter dorsal fin filament, lacks broad diffuse bars on side (may have narrower chevron-shaped ones), and body shape is slightly more slender; North West Cape northwards; Indo-W. Pacific; to 90 cm. ★★

466 BASSETT-HULLS TREVALLY
Uraspis uraspis (Günther)
Inhabits coastal waters; distinguished by white tongue and floor of mouth vividly contrasted against rest of mouth which is blue-black, also general colouration of body and fins dusky to black shading to grey on lower part of body; Port Hedland northwards; Indo-W. Pacific; to 30 cm; 1.4 kg. ★★

467 FLAGTAIL BLANQUILLO
Malacanthus brevirostris Guichenot
Inhabits sand and rubble areas in the vicinity of coral reefs, often occurs in pairs; distinguished by elongate shape with long-based dorsal and anal fins, and pair of dark stripes on tail; Ningaloo Reef northwards; Indo-C. Pacific; to 30 cm.

468 TAILOR
Pomatomus saltator (Linnaeus)
Inhabits estuaries and inshore waters; more common in southern waters; related to trevallies and king-fishes (412-463), but has larger, more obvious scales and very sharp teeth; colour silvery with green or bluish tinge; northward to Point Quobba; temperate and subtropical Atlantic and Indo-W. Pacific; to 120 cm; 12.1 kg. ★★★

469 COBIA
Rachycentron canadus (Linnaeus)
Inhabits coastal waters; distinguished by white stripe on side (may fade after death) and very small dorsal spines; also known as Black kingfish and Sergeant fish; Cape Naturaliste northwards; Atlantic and Indo-C. Pacific; to 202 cm; 61.5 kg. ★★★

470 SLENDER SUCKERFISH
Echeneis naucrates Linnaeus
Inhabits inshore reefs and offshore oceanic waters; distinguished by striped pattern and flattened head with sucking-disk structure on top; disc is used for attaching itself to larger fishes such as sharks, rays and mackerels; Albany northwards; worldwide temperate and tropical seas; to 100 cm; 2.381 kg. ★★★

471 REMORA
Remora remora (Linnaeus)
Inhabits coastal and oceanic waters; similar in shape and with same sucking-disk apparatus as 470, but lacks stripes and is overall brownish-black to grey in colour; entire coast of W.A., worldwide temperate and tropical seas; to 45 cm.

472 MOONFISH
Mene maculata (Bloch & Schneider)
Inhabits deeper coastal waters, sometimes enters estuaries; distinguished by oval shape, protrusible jaws, very low dorsal and anal fins, and silvery colour; Shark Bay northwards; Indo-W. Pacific; to 24 cm. ★★

473 TOOTHPONY
Gaza minuta (Bloch)
Inhabits coastal waters to 40 m depth; distinguished from other ponyfishes on this page by canine-like teeth in jaws; Exmouth Gulf northwards; Indo-W. Pacific; to 14 cm.

474 ORANGEFIN PONYFISH
Leiognathus bindus (Valenciennes)
Inhabits coastal waters to 40 m depth, occurs in schools; distinguished by protrusible jaws and black-edged orange tip of spiny dorsal fin; Shark Bay northwards; Indo-W. Pacific; to 11 cm.

475 COMMON PONYFISH
Leiognathus equulus Forsskål
Inhabits coastal waters, sometimes enters estuaries, occurs in schools; distinguished by protrusible jaws and thin vertical lines on back; *L. splendens* (not shown) is similar, but has a black tip on the spiny dorsal fin; Broome northwards; Indo-W. Pacific; to 24 cm.

476 SLENDER PONYFISH
Leiognathus elongatus (Günther)
Inhabits coastal waters to 40 m depth, occurs in schools; distinguished by protrusible jaws and very elongate shape; Shark Bay northwards; Indo-W. Pacific; to 12 cm.

477 WHIPFIN PONYFISH
Leiognathus leuciscus (Günther)
Inhabits coastal waters, occurs in schools; distinguished by protrusible jaws, elongate filament on spiny dorsal fin, and dark vermiculations on back; *L. moretoniensis* (not shown) has similar shape and markings, but has small scales on the cheek (versus no scales); *L. fasciatus* not shown is also similar, but has dark vertical lines on back and is deeper-bodied; Shark Bay northwards; Indo-W. Pacific; to 12 cm.

478 SMITHURST'S PONYFISH
Leiognathus smithursti (Ramsay & Ogilby)
Inhabits coastal waters to 40 m depth, occurs in schools; distinguished by protrusible jaws and elongate filaments at front of both dorsal and anal fins; Barrow Island northwards; Indo-W. Pacific; to 16 cm.

479 PUGNOSE PONYFISH
Secutor ruconis (Hamilton)
Inhabits coastal waters, sometimes entering estuaries and rivers, occurs in schools; distinguished by jaw that when protruded point upwards (those of other ponyfishes on this page point either slightly downwards or straight ahead), differs from *S. insidiator* (not shown) by having scales present on cheek; Exmouth Gulf northwards; Indo-W. Pacific; to 8 cm.

480 TRIPLETAIL
Lobotes surinamensis (Bloch)
Inhabits mangrove estuaries and lower reaches of freshwater streams; distinguished by steep, sloping humped forehead and large rounded posterior lobes of dorsal and anal fins that are about equal to the tail in size; Mandurah northwards; worldwide in tropical and subtropical seas; to 100 cm; 7.05 kg. ★★★★

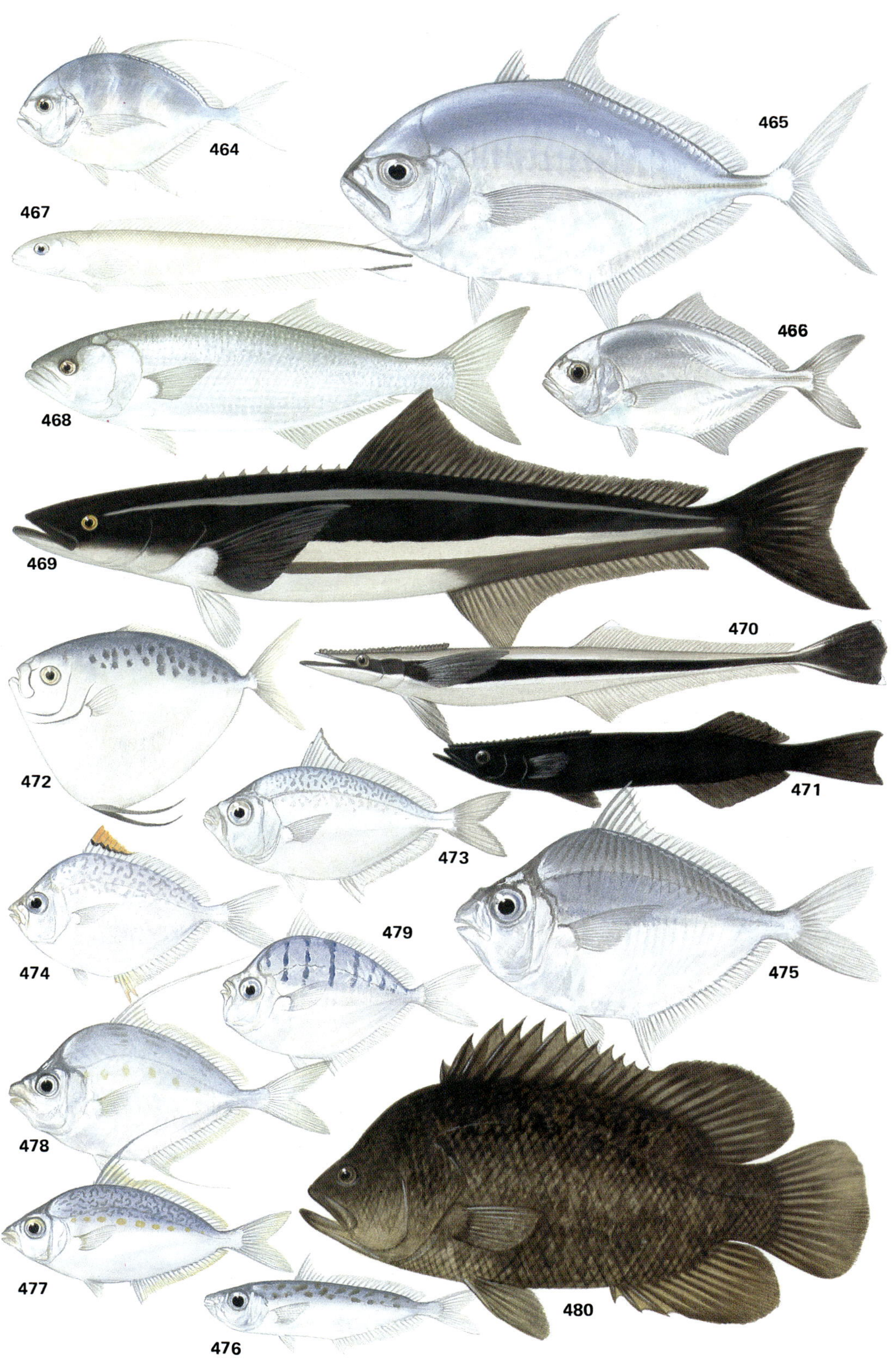

481 SMALL-TOOTHED JOBFISH
Aphareus rutilans Cuvier
Inhabits reefs and rocky bottoms to at least 100 m depth; distinguished by elongate body shape and deeply forked tail; Bernier Island northwards; Indo-C. Pacific; to 80 cm; 3.84 kg. ★★★

482 GREEN JOBFISH
Aprion virescens Valenciennes
Inhabits reef areas to at least 100 m depth; distinguished by dark green to blue-grey colour and dark patches along base of dorsal fin, snout blunter and tail less forked than 481. Ningaloo Reef northwards; Indo-C. Pacific; to 100 cm; 11.6 kg. ★★★

483 RUBY SNAPPER
Etelis carbunculus Cuvier
Inhabits rocky offshore reefs, usually between 90-300 m depth; distinguished from 484 by brighter red colouration and less forked tail; North West Shelf; Indo-C. Pacific; to 80 cm. ★★★

484 PALE SNAPPER
Etelis radiosus Anderson
Inhabits rocky offshore reefs, usually between 90-200 m depth; distinguished from 483 by pale (pinkish) colour and more forked tail; North West Shelf; W. Pacific and Indian Ocean; to 60 cm. ★★★

485 TANG'S SNAPPER
Lipocheilus carnolabrum (Chan)
Inhabits rocky bottoms of the continental shelf between 90-300 m depth; distinguished by thick fleshy protrusion at the front of the upper lip; Dirk Hartog Island northwards; Indo-W. Pacific; to 60 cm. ★★★

486 ROSY SNAPPER
Pristipomoides filamentosus (Valenciennes)
Inhabits rocky bottoms, mainly between 90-360 m depth; distinguished by rosy colour and elongate tips at rear of dorsal and anal fins; Point Quobba northwards; Indo-W. Pacific; to 80 cm; 5.852 kg. ★★★★

487 SHARPTOOTH JOBFISH
Pristipomoides typus Bleeker
Inhabits rocky bottoms between about 40-100 m depth; similar to 488, but lacks orange stripes below eye; North West Cape northwards; mainly western Pacific; to 70 cm. ★★★

488 GOLD-BANDED JOBFISH
Pristipomoides multidens (Day)
Inhabits rocky bottoms between 40-200 m depth; similar to 487, but has pair of orange stripes below eye; North West Shelf; Indo-W. Pacific; to 90 cm; 5.852 kg. ★★★

489 CHINAMAN FISH
Symphorus nematophorus (Bleeker)
Inhabits inshore coral reefs and deeper offshore areas to at least 50 m; distinguished by distinctive shape and reddish colour of adults, young have elongate filaments on the rear part of the dorsal fin; Shark Bay northwards; mainly western Pacific; to 80 cm. Considered dangerous to eat due to its susceptibility to ciguatera. 17.59 kg.

490 MANGROVE JACK
Lutjanus argentimaculatus (Forsskål)
Inhabits inshore and offshore reefs to 100 m depth, young and subadults found in mangrove estuaries; distinguished from 492 by taller dorsal fin, lack of stripes on side, and absence of black on fins, young have pronounced bars; Dampier Archipelago northwards; Indo-W. Pacific; to 120 cm; 11.2 kg. ★★★

491 INDONESIAN SEAPERCH
Lutjanus bitaeniatus (Valenciennes)
Inhabits deeper offshore reefs between about 40-70 m; distinguished by reddish colouration and absence of stripes, although juvenile has a black stripe along middle of sides; North West Shelf; mainly Indonesia; to 30 cm. ★★★

492 RED BASS
Lutjanus bohar (Forsskål)
Inhabits coral reefs to at least 70 m depth; distinguished by reddish hue, faint stripes on lower side, and blackish colour on dorsal and anal fins (also tip of pelvics and upper edge of pectorals), juveniles and subadults with white spot below rear part of dorsal fin; Ningaloo Reef northwards; Indo-C. Pacific; to 75 cm. Reported to be good eating, but Australian specimens are best avoided due to the possibility of ciguatera poisoning. 11.2 kg.

FISH POISONING

Although relatively few cases of a form of fish poisoning known as Ciguatera have been reported from Australia, those of us, particularly anglers, who are regular consumers of fresh seafood should be aware of its dangers. This topic is included here because two of the snappers shown on this plate, nos. 489 and 492 have been frequently implicated in various regions of the Indo-Pacific. The incidence of poisonous fishes is alarmingly high in some areas, particularly around some Polynesian islands. The exact nature of the causative agent has long puzzled scientists. The same species of fish can be poisonous at one locality, but safe to eat from a nearby reef. It is now believed that the poisonous properties are caused by a toxic dinoflagellate *(Gambierdiscus toxicus)*, that lives on dead coral or among benthic algae and is first consumed by herbivorous fishes who are eventually eaten by larger predatory fishes. The toxin is apparently accumulative and the largest fishes are potentially the most dangerous. In fact at some locations known to harbour ciguatoxic fishes the young and small adults of a particular species may be eaten with impunity while large adults may be toxic.

The symptoms from eating ciguatoxic fish appear from one to 10 hours later and range from mild dizziness, diarrhoea, and a numb sensation of the lips, hands and fingers to extreme nausea, coma and total respiratory failure causing death. The degree of poisoning depends on the amount of fish that is consumed and the concentration of toxin it contains.

481
482
483
484
485
486
487
488
489
490
juv.
juv.
491
juv.
492
juv.

493 CRIMSON SEAPERCH
Lutjanus erythropterus Bloch
Inhabits trawling grounds and reefs to depths of at least 100 m; similar to 496, but head and mouth much smaller; also known as Saddle-tailed seaperch; Dampier Archipelago northwards; Indo-W. Pacific; to 100 cm; 2.51 kg. ★★★

494 STRIPEY SEAPERCH
Lutjanus carponotatus (Richardson)
Inhabits coral reefs in sheltered lagoons and outer reef areas; distinguished by striped pattern; Shark Bay northwards; mainly Indo-Australian Archipelago; to 40 cm; 1.375 kg. ★★★

495 CHECKERED SEAPERCH
Lutjanus decussatus (Cuvier)
Inhabits coral reefs to at least 30 m depth; distinguished by checkered pattern on the upper sides and black spot at the base of the tail; Dampier Archipelago northwards; mainly Indo-Australian Archipelago; to 30 cm.

496 SADDLE-TAILED SEAPERCH
Lutjanus malabaricus Schneider
Inhabits coastal and offshore reefs, also flat-bottom trawling grounds; similar to 493, but head and mouth much larger; also known as Scarlet seaperch; Shark Bay northwards; Indo-W. Pacific; to 100 cm; 6.3 kg. ★★★

497 MOLUCCAN SEAPERCH
Lutjanus boutton (Lacepéde)
Inhabits offshore coral reefs to depths of 50 m; distinguished by series of faint yellow stripes on the sides and a deep notch in the rear margin of the cheek; North West Shelf reefs; mainly W. Pacific; to 30 cm. ★★★

498 BLUESTRIPE SEAPERCH
Lutjanus kasmira (Forsskål)
Inhabits inshore coral reefs, but rare on our coast; distinguished from the more common 499 by its white belly and lack of a fifth blue stripe; Shark Bay northwards; Indo-C. Pacific; to 35 cm. ★★★

499 FIVE-LINED SEAPERCH
Lutjanus quinquelineatus Bloch
Inhabits sheltered lagoon reefs and outer reef areas; similar to 498, but has extra blue stripe on belly; also known as Blue-banded seaperch; Shark Bay northwards; Indo-W. Pacific; to 38 cm; .26 kg. ★★★

500 BLACK-SPOT SEAPERCH
Lutjanus fulviflamma (Forsskål)
Inhabits coral reefs to at least 35 depth; distinguished by yellow stripes on sides and elongate black spot on back, snout blunter than 504; Shark Bay northwards; Indo-C. Pacific; to 35 cm; .33 kg. ★★★

501 MAORI SEAPERCH
Lutjanus rivulatus (Cuvier)
Inhabits inshore coral reefs and also deeper offshore waters; distinguished by "blubbery" lips and wavy lines on head; North West Shelf reefs; Indo-C. Pacific; to 65 cm; 4 kg. ★★★

502 RED EMPEROR
Lutjanus sebae (Cuvier)
Inhabits the vicinity of coral reefs, often over adjacent sand and rubble flats; distinguished by shape and overall red-pink colour, juveniles and subadults have distinctive pattern of dark bars; Shark Bay northwards; Indo-W. Pacific; to 100 cm; 16 kg. ★★★

503 STRIPED SEAPERCH
Lutjanus vitta (Quoy & Gaimard)
Inhabits the vicinity of coral reefs, also flat bottoms with coral outcrops, sponge and sea whips; distinguished by brown or blackish stripe along middle of sides; Abrolhos northwards; Indo-W. Pacific; to 40 cm. ★★★

504 MOSES PERCH
Lutjanus russelli (Bleeker)
Inhabits inshore rock or coral reefs, also deeper offshore reefs to at least 80 m depth; distinguished by reddish colouration and black spot (sometimes faint) on back, has more pointed snout than 500; Shark Bay northwards; Indo-W. Pacific; to 45 cm; 2.22 kg. ★★★

505 BIGEYE SEAPERCH
Lutjanus lutjanus Bloch
Inhabits offshore coral reefs and trawling grounds to at least 90 m depth; distinguished from 503 by lighter midlateral stripe and much narrower space between eye and upper jaw, formerly known as *L. lineolatus;* Ningaloo Reef northwards; Indo-W. Pacific; to 30 cm; .1 kg. ★★★

506 DARK-TAILED SEAPERCH
Lutjanus lemniscatus Valenciennes
Inhabits inshore coral reefs and deeper offshore reefs; distinguished by blackish tail; Point Quobba northwards; mainly Indo-Australian Archipelago; to 65 cm; 2.4 kg. ★★★

TROPICAL SNAPPERS

The members of the family Lutjanidae are generally known worldwide as snappers. However, this name sometimes causes confusion in Australia, because it is also used for at least two other common fishes belonging to two different families. The common Snapper of Australia's southern half belongs to the family Sparidae and is perhaps the best known of the two. The other fish is a member of the family Lethrinidae (Plate 33). Although it is officially known as the Spangled emperor (528), it is often referred to as North West snapper by W.A. anglers.

Snappers and seaperches featured on Plates 30-31 are primarily inhabitants of tropical reefs. Most species are distributed in the Indo-Pacific region, although they also occur in the Atlantic Ocean. *Lutjanus* (490-506) is by far the largest genus, containing 65 species, many of which are brightly coloured. They are active nocturnal predators that feed mainly on fishes, but crabs, shrimps, gastropods, cephalopods, and planktonic organisms are also eaten. The larger, deep-bodied snappers (*Lutjanus* for example) usually have well developed canine teeth, adapted for seizing and holding larger prey items. More slender snappers such as *Pristipomoides* (486-488) and *Etelis* (483-484) have weaker dentition and consume a significant amount of plankton. The larger species of *Lutjanus*, particularly the 'red snappers' are favourite angling fishes and also commercially important.

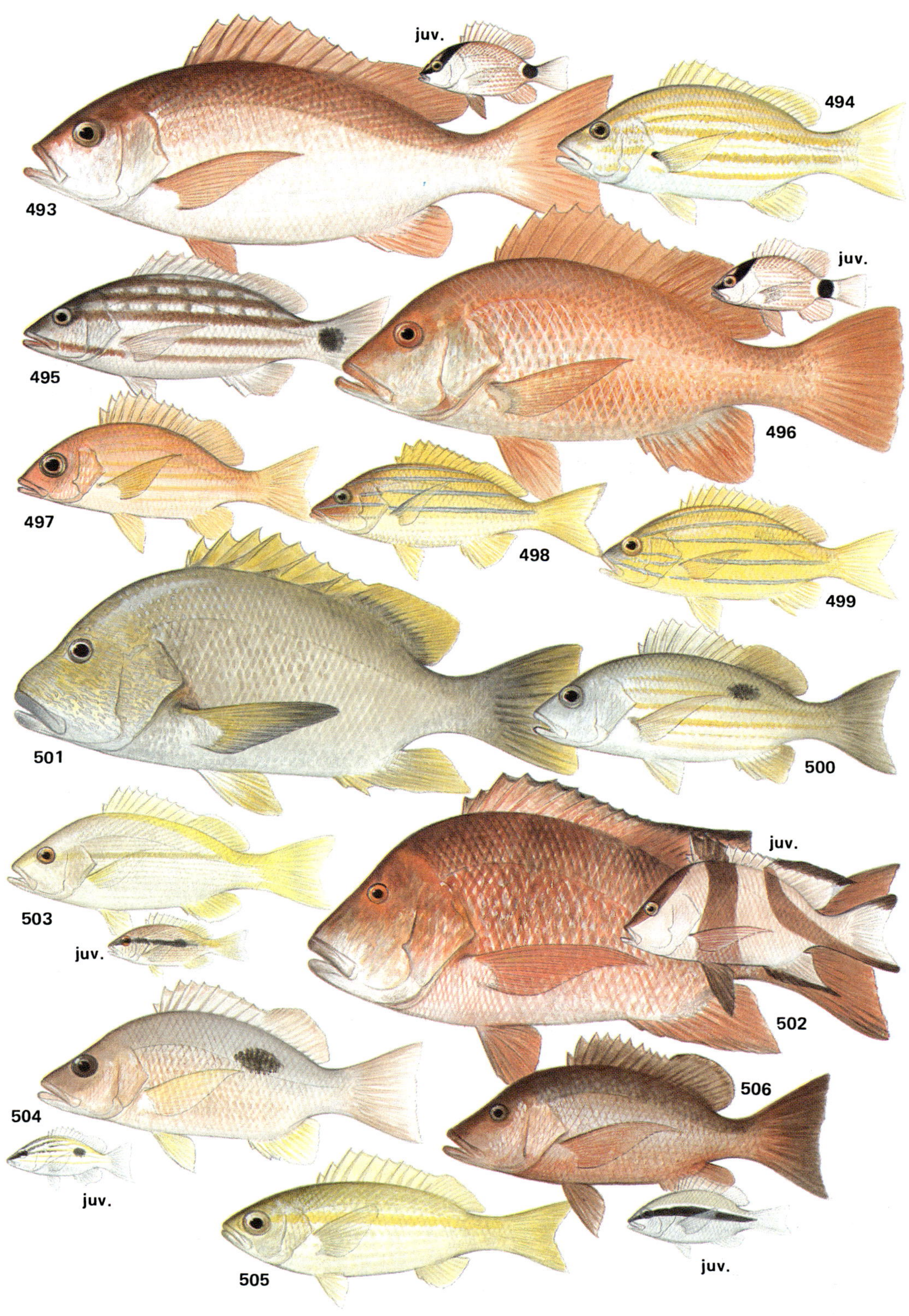

juv.
493
494
495
496
juv.
497
498
499
501
500
503
juv.
502
juv.
504
506
juv.
505
juv.

507 MINSTREL SWEETLIPS
Plectorhinchus schotaf (Forsskål)
Inhabits caves and crevices of coral reefs; similar to 511 but distinguished by 11 or 12 dorsal spines (versus 13 or 14); Kalbarri northwards; Indo-W. Pacific; to 60 cm; 4.167 kg. ★★

508 MANY-SPOTTED SWEETLIPS
Pectorhinchus chaetodontoides (Lacepède)
Inhabits caves and crevices of coral reefs; distinguished by spotted pattern, juvenile has peculiar undulating motion; Coral Bay northwards; Indo-W. Pacific; to 60 cm; 7.0 kg. ★★

509 GOLD-SPOTTED SWEETLIPS
Plectorhinchus flavomaculatus (Ehrenberg)
See Plate 70.

510 MANY-LINED SWEETLIPS
Plectorhinchus multivittatum (Macleay)
Inhabits coral reefs; distinguished from 509 (p. 158) by yellow fins; Point Quobba northwards; Indo-W. Pacific; to 50 cm. ★★

510a CELEBES SWEETLIPS
Plectorhinchus celebicus Bleeker
Inhabits coral reefs; distinguished by yellow stripes on head and sides; Shark Bay northwards; W. Pacific; to 40 cm. ★★

511 BROWN SWEETLIPS
Plectorhinchus gibbosus (Lacepède)
Inhabits coral reefs; similar to 507 but distinguished by 13 or 14 dorsal spines (versus 11 or 12); formerly known as *P. nigrus;* Point Quobba northwards; Indo-W. Pacific; to 60 cm; 5.7 kg. ★★

512 PAINTED SWEETLIPS
Diagramma pictum (Thunberg)
Inhabits coral reefs, sometimes seen in schools; adults distinguished by silvery body colour, smaller specimens have spots or stripes, but differs from 508 in having 9 or 10 dorsal spines (versus 12) and lacks notch in dorsal fin profile; Rottnest Island northwards; Indo-W. Pacific; to 90 cm; 6.258 kg. ★★

513 RIBBON SWEETLIPS
Plectorhinchus polytaenia (Bleeker)
Inhabits coral reefs; distinguished by vivid pattern of dark-edged orange and white stripes; Onslow northwards; Indo-Australian Archipelago; to 40 cm. ★★

514 SPOTTED JAVELINFISH
Pomadasys kaakan (Cuvier)
Inhabits estuaries and coastal waters usually over sand or mud bottoms; distinguished by series of spots arranged in vertical rows on upper side and spots on dorsal fin; formerly known as *P. hasta;* Shark Bay northwards; Indo-W. Pacific; to 38 cm; 6.2 kg. ★★★

515 BLOTCHED JAVELINFISH
Pomadasys maculatum (Bloch)
Inhabits estuaries and coastal waters usually over sand or mud bottoms; distinguished by large dark blotches mainly on upper sides; Shark Bay northwards; Indo-W. Pacific; to 45 cm. ★★★

516 LINED JAVELINFISH
Hapalogenys kishinouyei Smith & Pope
Inhabits coastal waters, usually over mud or sand bottoms; distinguished by brown stripes on upper part of back; Broome northwards; Indo-W. Pacific; to 30 cm. ★★★

517 RED-BELLIED FUSILER
Caesio cuning Cuvier
Inhabits coral reefs, forming midwater schools; distinguished by yellow tail, similar to 518 but lacks black tips of tail; Coral Bay northwards, Indo-W. Pacific; to 43 cm; .812 kg. ★★★

518 BLUE FUSILER
Caesio lunaris Cuvier
Inhabits coral reefs, forming midwater schools; similar to 517 but has blue tail with black tips; Coral Bay northwards; Indo-W. Pacific; to 38 cm. ★★★

519 BLACK-TIPPED FUSILER
Pterocaesio diagramma (Bleeker)
Inhabits coral reefs, forming midwater schools; distinguished from 517 and 518 by more slender shape and pair of yellow stripes on upper side; to 28 cm; .38 kg. ★★

520 BANJOFISH
Banjos banjos (Richardson)
Inhabits coastal waters usually over sand bottoms; distinguished by stout dorsal and anal spines, angular forehead and black margin on soft dorsal and caudal fins; Abrolhos northwards; Indo-W. Pacific; to 30 cm.

SWEETLIPS, ETC.

Most of the fishes on Plate 32 belong to the family Haemulidae, generally known on a worldwide basis as Grunts. The most common representatives on northern Australian reefs are in the genus *Plectorhinchus* (507-514) and are referred to as Sweetlips. Perhaps the most notable aspect of these fishes, besides their use as food, are the dramatic changes which many of the species undergo. Juveniles often exhibit striking patterns that gradually change with advancing age. Two of these transformations are shown for 508 and 512. The young of *Plectorhinchus chaetodontoides* is thought to mimic unpalatable soft-bodied invertebrates such as nudibranchs or turbellarians, and thus enjoys some measure of freedom from predators. Young grunts feed on zooplankton; whereas adults eat a variety of benthic invertebrates. Maximum size is about 1 m TL, but many species are under 40-50 cm. Worldwide the family contains about 175 species and is represented in all tropical seas.

Fusilers (517-519) are closely related to Snappers (Plates 30-31). The family (Caesionidae) contains 20 species which are distributed across the Indo-Pacific region. They frequently form schools containing up to several hundred individuals that feed in midwater on plankton above coral reefs.

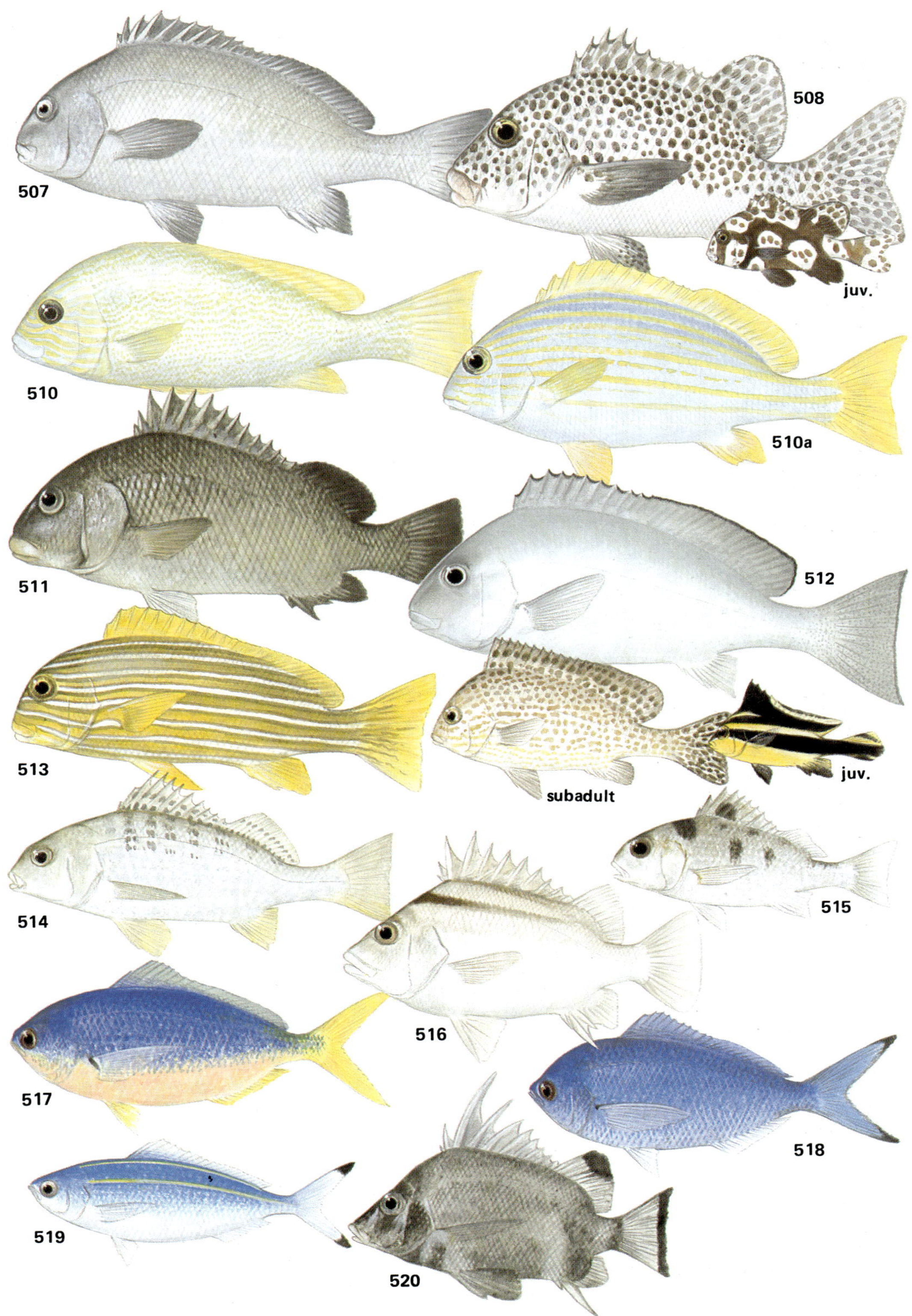

507
508
juv.
510
510a
511
512
513
subadult
juv.
514
515
516
517
518
519
520

521 HUMPNOSE BIG-EYE BREAM
Monotaxis grandoculus (Forsskål)
Inhabits coral reefs, frequently over adjacent sand or rubble areas; distinguished by large eye and bluntly rounded snout, juvenile with dark bars that are reduced to saddles on back of adults; Ningaloo Reef northwards; Indo-C. Pacific; to 60 cm. ★★★

522 SWALLOWTAIL SEABREAM
Gymnocranius elongatus Senta
Inhabits sand bottoms and trawling grounds; distinguished by 5 narrow bars on side and strongly forked tail; Onslow northwards; E. Indian Ocean and W. Pacific; to 18 cm. ★★

523 NAKED-HEADED SEABREAM
Gymnocranius griseus (Schlegel)
Inhabits sand bottoms and trawling grounds; similar to 522 but bars generally wider and more diffuse and tail not as forked; Point Quobba northwards; Indo-W. Pacific; to 80 cm. ★★

524 ROBINSON'S SEABREAM
Gymnocranius robinsoni (Gilchrist & Thompson)
Inhabits sand bottoms and trawling grounds; distinguished by wavy blue lines on cheek and snout; also known as Blue-lined seabream; Fremantle northwards; Indo-W. Pacific; to 80 cm; 5.167 kg. ★★★

525 LONG-NOSED EMPEROR
Lethrinus elongatus (Valenciennes)
Inhabits coral reefs; distinguished by long painted snout and overall greyish colour; Ningaloo Reef northwards; Indo-C. Pacific; to 100 cm; 9.6 kg. ★★★

526 SWEETLIP EMPEROR
Lethrinus miniatus (Bloch & Schneider)
Inhabits coral reefs; distinguished by orange area around eyes, bright red dorsal fin, and red patch at base of pectoral fin; sometimes a series of dark bars on side; formerly known as *L. chrysostomus;* Perth area northwards, but rare south of Abrolhos; N. Australia and S.W. Pacific; to 90 cm; 6.776 kg. ★★★★

527 BLUE-SPOTTED EMPEROR
Lethrinus laticaudis Alleyne & Macleay
Inhabits coral reefs and adjacent sand-rubble areas; similar to 528, but has blue spots (versus blue bars) on cheek, whitish oblique bar behind eye, and brownish band from eye to mouth; often assumes barred pattern when under stress; Ningaloo Reef northwards; E. Indian Ocean and W. Pacific; to 80 cm; 1.55 kg. ★★★★

528 SPANGLED EMPEROR
Lethrinus nebulosus (Forsskål)
Inhabits coral reefs, usually over adjacent sandy areas; distinguished by blue spots on scales, similar to 527, but has blue bars (versus spots) on cheek and lacks other marks on head as described above; also known as North-West snapper; Rottnest Island northwards; Indo-W. Pacific; to 86 cm; 6.577 kg. ★★★★

529 BLUE-LINED EMPEROR
Lethrinus sp.
Inhabits coastal reefs, usually over adjacent sand or rubble areas; similar to 528, but has dark streak on each scale instead of blue spots, and has short blue lines radiating from eye to which several cross forehead and connect with opposite eye; Ningaloo Reef northwards; mainly W. Pacific; to 60 cm; 1.912 kg. ★★★★

530 YELLOW-TAILED EMPEROR
Lethrinus atkinsoni Seale
See Plate 70.

531 THREADFIN EMPEROR
Lethrinus nematacanthus Bleeker
Inhabits sand-weed areas, sometimes in estuaries; distinguished by dark blotch above pectoral fin and elongate filament near front of dorsal fin; also known as Longspine emperor and Lancer; Fremantle northwards, but rare south of Shark Bay; E; Indian Ocean and W. Pacific; to 25 cm; .429 kg. ★★

532 PURPLE-HEADED EMPEROR
Lethrinus lentjan (Lacepède)
Inhabits sandy areas next to coral reefs; distinguished by pale spots on scales, red margin on gill cover, and red spot at base of pectoral fin; head of fresh caught specimens usually turns to purple colour; also known as Pink-eared emperor; Broome northwards; Indo-W. Pacific; to 50 cm; 1.35 kg. ★★★

533 SPOTCHEEK EMPEROR
Lethrinus rubrioperculatus Sato
Inhabits coral reefs and trawling grounds; distinguished by elongate shape and red or brown blotch on upper margin of gill cover; colour varies from uniform greyish to blotchy pattern as shown (most emperors can quickly assume a similar pattern); *L. semicinctus* (not shown) is similar but has a downward slanting dark blotch below posterior part of dorsal fin; Ningaloo Reef northwards; Indo-C. Pacific; to 50 cm; .62 kg. ★★★

534 VARIEGATED EMPEROR
Lethrinus variegatus Valenciennes
Inhabits sand-weed areas near coral reefs; distinguished by elongate shape and blotchy pattern with faint cross bands on tail, similar to 533, but lacks blotch on upper gill cover; Ningaloo Reef northwards; Indo-W. Pacific; to 20 cm. ★★

535 NORTH WEST BLACK BREAM
Acanthopagrus palmaris (Whitley)
Inhabits coastal reefs, sometimes in estuaries; similar to 536, but is generally darker and lacks yellow on fins; Shark Bay northwards; N.W. Australia only; to 40 cm. ★★★

536 WESTERN YELLOWFIN BREAM
Acanthopagrus latus (Houttuyn)
Inhabits coastal reefs, frequently in schools; similar to 535, but lighter in colour and with yellow fins; Shark Bay northwards; N. Indian Ocean and W. Pacific; to 45 cm; 1.975 kg. ★★★

537 LONG-SPINED SNAPPER
Argyrops spinifer (Forsskål)
Inhabits coastal waters and deeper trawling grounds; distinguished by filamentous dorsal spines and angular forehead profile; Shark Bay northwards; Indo-W. Pacific; to 65 cm; 2.0 kg. ★★★

538 DEEPSEA SNAPPER
Dentex tumifrons (Temminck & Schlegel)
Inhabits trawling grounds; similar to 537, but lacks filamentous dorsal spines and forehead is more rounded; North West Cape northwards; mainly W. Pacific; to 35 cm. ★★★

521
juv.
522
523
524
525
526
527
528
527
529
531
juv.
532
juv.
533
534
537
536
538
535

539 BALI THREADFIN-BREAM
Nemipterus balinensis (Bleeker)
Inhabits trawling grounds; distinguished by single yellow band on side and slender shape; not yet recorded from N.W. Australia, but may occur there; Indo-Malayan Archipelago; to 24 cm. ★★

540 YELLOWBELLY THREADFIN-BREAM
Nemipterus bathybius Snyder
Inhabits trawling grounds; distinguished by 2 yellow bands on side and 2 yellow stripes along ventral profile of head and body; Barrow Island northwards; W. Pacific and E. Indian Ocean; to 28 cm. ★★

541 FIVE-LINED THREADFIN-BREAM
Nemipterus celebicus (Bleeker)
Inhabits trawling grounds; distinguished by 4-5 silvery yellow stripes on side, equal sized tail lobes and reddish tip on upper lobe of tail; North West Shelf; Indo-Australian Archipelago; to 22 cm. ★★

542 ROSY THREADFIN-BREAM
Nemipterus furcosus (Valenciennes)
Inhabits trawling grounds; distinguished by series of faint dark blotches on upper back and uniform fins without stripes or spots; North West Shelf; W. Pacific and E. Indian Ocean; to 35 cm; .45 kg. ★★

543 ORNATE THREADFIN-BREAM
Nemipterus hexodon (Quoy & Gaimard)
Inhabits trawling grounds; distinguished by broad yellow stripe on upper side (with narrower stripes below), oblique yellow stripe on dorsal fin and, yellow tip on upper lobe of tail; Exmouth Gulf northwards; Indo-Australian Archipelago and S.E. Asia; to 30 cm. ★★

544 TWIN-LINED THREADFIN-BREAM
Nemipterus isacanthus (Bleeker)
Inhabits trawling grounds; distinguished by 2 broad yellow stripes along side, bright yellow band along ventral profile, and yellow patch below eye; North West Shelf; Indo-Australian Archipelago and S.E. Asia; to 24 cm. ★★

545 JAPANESE THREADFIN-BREAM
Nemipterus japonicus (Bloch)
Inhabits trawling grounds; distinguished by broad yellow band along base of dorsal fin, bright pink-red "sholder" spot, yellow stripe along ventral profile, and elongate yellow filament on upper lobe of tail; colour generally more pink (less blue than shown); North West Shelf; W. Pacific and E. Indian Ocean; to 32 cm. ★★

546 SLENDER YELLOW-TIPPED THREADFIN-BREAM
Nemipterus mesoprion (Bleeker)
Inhabits trawling grounds; similar to 544, but lacks yellow patch below eye; North West Shelf; mainly Indo-Australian Archipelago; to 20 cm. ★★

547 YELLOW-CHEEKED THREADFIN-BREAM
Nemipterus zysron (Bleeker)
Inhabits trawling grounds; distinguished by broad yellow stripe below eye and elongate yellow filament on upper lobe of tail; Port Hedland northwards; Indo-Australian Archipelago and Andaman Sea; to 28 cm. ★★

548 DOUBLEWHIP THREADFIN-BREAM
Nemipterus nematophorus (Bleeker)
Inhabits trawling grounds; distinguished by elongate filaments at front of dorsal fin and on upper lobe of tail (these features not shown, but usually present); not yet recorded from N.W. Australia, but may occur there; Indo-Australian Archipelago; to 25 cm. ★★★

549 NOTCHED THREADFIN-BREAM
Nemipterus peronii (Valenciennes)
Inhabits trawling grounds; distinguished by deep notches along margin of dorsal fin between spines; Port Hedland northwards; Indo-W. Pacific; to 30 cm. ★★

550 PINKFIN THREADFIN-BREAM
Nemipterus sp.
Inhabits trawling grounds; distinguished by 3 yellow bands between eye and upper lip, 6-7 yellow stripes on side, a yellow margin on dorsal fin and single yellow stripe on anal fin; not yet recorded from N. W. Australia, but may occur there; Indo-Australian Archipelago; to 30 cm. ★★

551 YELLOW-LIPPED THREADFIN-BREAM
Nemipterus virgatus (Houttuyn)
Inhabits trawling grounds; distinguished by bright yellow lips (not shown), thin yellow stripes on sides; pair of thin stripes on dorsal and anal fins, and West Shelf; W. Pacific and E. Indian Ocean; to 40 cm. ★★

552 YELLOW-TIPPED THREADFIN-BREAM
Nemipterus nematopus (Bleeker)
Inhabits trawling grounds; distinguished by pair of yellow stripes on side, lowermost wider and curved slightly upwards above base of pectoral fin, also has vivid yellow spot on tip of upper tail lobe and frequently with pair of yellow stripes along ventral surface of head and body; North West Shelf; W. Pacific and E. Indian Ocean; to 40 cm. ★★

THREADFIN AND MONOCLE BREAMS

The Sea Breams of the family Nemipteridae (Plates 34 and 35; nos. 539-564) are confined to the Indo-west Pacific region, primarily in tropical and subtropical latitudes. The family contains about 45 species, several of which are still undescribed. The largest genera are *Nemipterus* (539-552) and *Scolopsis* (559-563), each having about 15-18 species. Both are commonly offered in Southeast Asian fish markets, although they are relatively small in size (usually less than 40 cm TL). *Scolopsis* and *Pentapodus* (555-558), are closely associated with coral or rubble reefs, being particularly abundant in sandy areas between reefs. *Nemipterus* and *Parascolopsis* (553-554) generally occur in deeper water, frequently between 30 and 100 m depth. They form a significant portion of trawl catches on continental shelves. Breams forage for worms, crustaceans, molluscs, and other invertebrates; small fishes are also sometimes eaten.

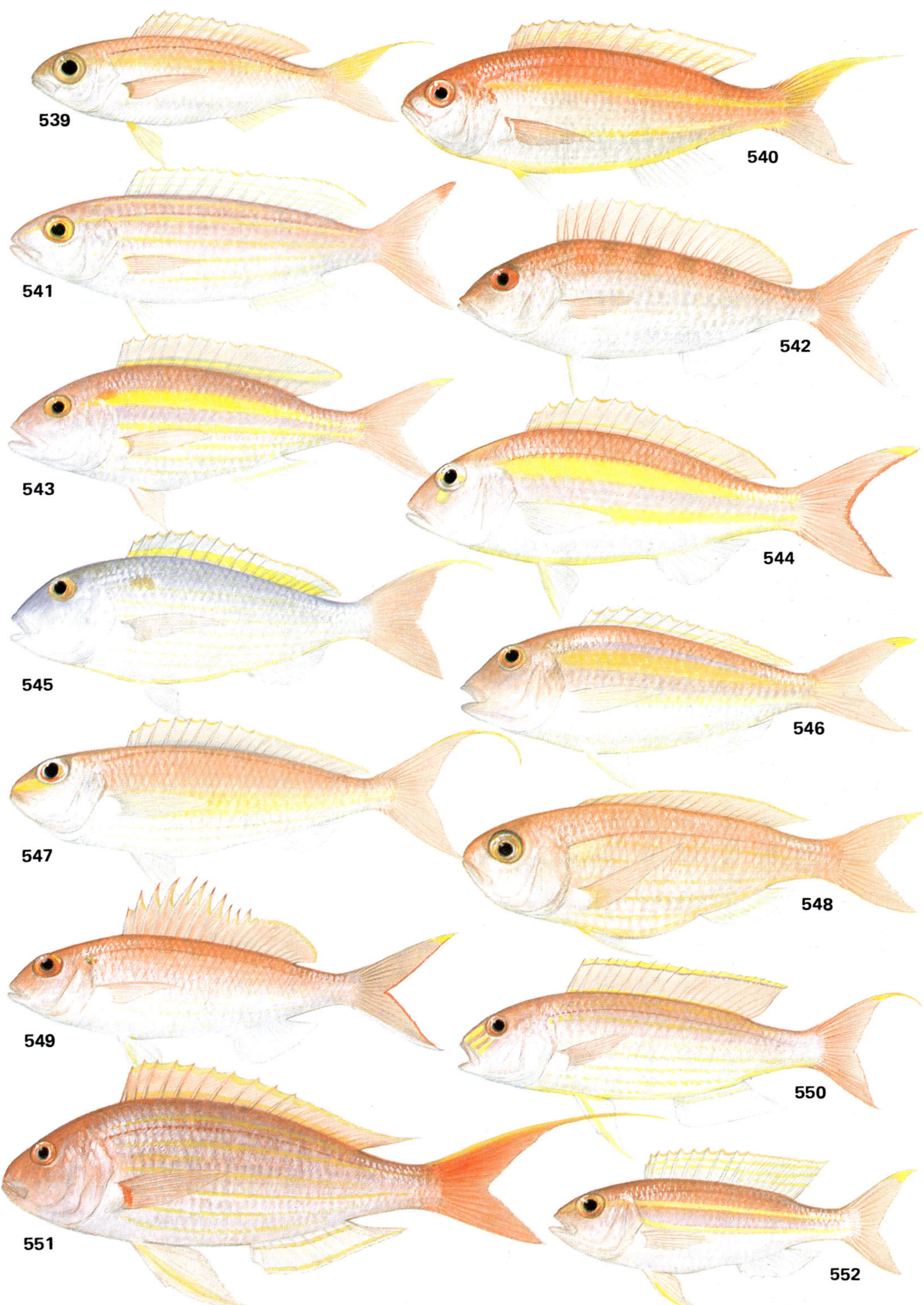

553 ROSY DWARF MONOCLE BREAM
Parascolopsis eriomma Jordan & Richardson
Inhabits offshore trawling grounds; distinguished by rosy colour on back with diffuse yellow longitudinal band along middle of side; Onslow northwards; Indian Ocean and Indo-Australian Archipelago; to 30 cm. ★★

554 SADDLED DWARF MONOCLE BREAM
Parascolopsis tanyactis Russell
Inhabits offshore trawling grounds in 45-80 m depth; distinguished by 2 dark bars on back, becoming diffuse on lower side and additional bar across tail base, also 4th and 5th soft dorsal rays elongated (not shown); *P. rufomaculatus* (not shown) is similar, but soft dorsal rays not elongated and has broad golden stripe along middle of side; Carnarvon northwards; Indo-Australian Archipelago; to 20 cm. ★★

555 PURPLE THREADFIN-BREAM
Pentapodus emeryii (Richardson)
Inhabits coastal reefs; distinguished by bright yellow stripe from eye to tail, adults more bluish than shown and have long filaments at the tip of each caudal lobe; Ningaloo Reef northwards; N. Australia only; to 25 cm; .807 kg. ★★

556 JAPANESE BUTTERFISH
Pentapodus nagasakiensis (Tanaka)
Inhabits sand rubble areas adjacent to deeper offshore reefs; distinguished by orange stripe from eye to tail base with pearly stripe just below it and often a narrower pale stripe along base of dorsal fin; North West Cape northwards; mainly W. Pacific; to 20 cm. ★★

557 FALSE WHIPTAIL
Pentapodus porosus (Valenciennes)
Inhabits sand and rubble bottoms frequently near coastal reefs; distinguished by broad longitudinal brown band on middle of sides and <-shaped blue mark and small black spot at tail base; Exmouth Gulf northwards; N.W. Australia only; to 25 cm; .117 kg. ★★

558 WESTERN BUTTERFISH
Pentapodus vitta Quoy & Gaimard
Inhabits sand-weed areas or rocky bottoms close to shore, often in schools; distinguished by dark stripe bordered by blue from snout to tail base; Geographe Bay to Dampier Archipelago; W.A. only; to 31 cm; .407 kg. ★★

559 MONOCLE BREAM
Scolopsis monogramma (Kuhl & van Hasselt)
Inhabits sandy areas in the vicinity of coral reefs; distinguished by 3 blue stripes on snout (often with orange between them), generally pale body except may have series of slanting dotted lines on middle of side which sometimes form a solid broad stripe when viewed underwater, and blue edged tail with prolonged filament on upper lobe; often misidentified as *S. temporalis;* also known as Barred-face spinecheek; Ningaloo Reef northwards; Andaman Sea and Indo-Australian Archipelago; to 30 cm; .25 kg. ★★

560 BRIDLED MONOCLE BREAM
Scolopsis bilineatus (Bloch)
Inhabits coral reefs; distinguished by curved black-edged white stripe from below eye to midbase of dorsal fin; also known as Bridled spinecheek and Double-lined monocle bream; Point Quobba

northwards; E. Indian Ocean and W. Pacific; to 23 cm. ★★

561 WHITECHEEK MONOCLE BREAM
Scolopsis vosmeri (Bloch)
Inhabits coastal reefs; distinguished by relatively deep body and white bar behind eye; Broome northwards; Indo-W. Pacific; to 25 cm. ★★

562 REDSPOT MONOCLE BREAM
Scolopsis taeniopterus (Kuhl & van Hasselt)
Inhabits sandy areas in the vicinity of coral reefs; similar to 559 , but has small red spot in axil ("armpit") of pectoral fin, lacks distinct blue stripes on snout, but has blue bands between eyes, and lacks dark spots or stripes on side; Exmouth Gulf northwards; Indo-Australian Archipelago; to 23 cm. ★★

563 PEARL-STREAKED MONOCLE BREAM
Scolopsis xenochrous Günther
Inhabits sand or rubble areas near coral reefs; distinguished by diagonal pearl-blue streak bordered with black spots above pectoral fin, series of slanted rows of dots on middle of side followed by broad white streak; Ningaloo Reef northwards; Andaman Sea and Indo-Australian Archipelago; to 18 cm. ★★

564 CORAL MONOCLE BREAM
Scaevius milii (Bory)
Inhabits coastal waters, frequently on sand or rubble bottoms; distinguished by 2 blue stripes on upper side ending on top of tail base, and wider orange stripe below along middle of side with small pale-edged black spot on upper tail base; also known as Jurgen; Fremantle to Dampier; W.A. only; to 25 cm; .24 kg. ★★

565 LONG-FINNED SILVER BIDDY
Pentaprion longimanus (Cantor)
Inhabits coastal waters over sand bottoms; similar to 566-568, but has much larger anal fin base (anal fin contains 12-14 soft rays versus 7-8 in others); Dampier northwards; Indo-W. Pacific; to 15 cm. ★

566 WHIPFIN SILVER-BIDDY
Gerres filamentosus Cuvier
Inhabits coastal waters including estuaries, usually on sand or mud bottoms; distinguished by long filament at front of dorsal fin and series of spots in vertical rows on side; Onslow northwards; Indo-W. Pacific; to 25 cm. ★

567 COMMON SILVER BIDDY
Gerres oyena (Forsskål)
Inhabits coastal waters including estuaries, usually on sand or mud bottoms; distinguished by relatively elongate body and lack of markings; also known as Silverbelly; Ningaloo Reef northwards; Indo-W. Pacific; to 25 cm. ★

568 ROACH
Gerres subfasciatus Cuvier
Inhabits coastal waters entering bays and estuaries, usually on sand or mud bottoms; distinguished by narrow dark bars on side; also known as Banded silver-biddy; Rottnest Island northwards; Australia only; 22 cm. ★

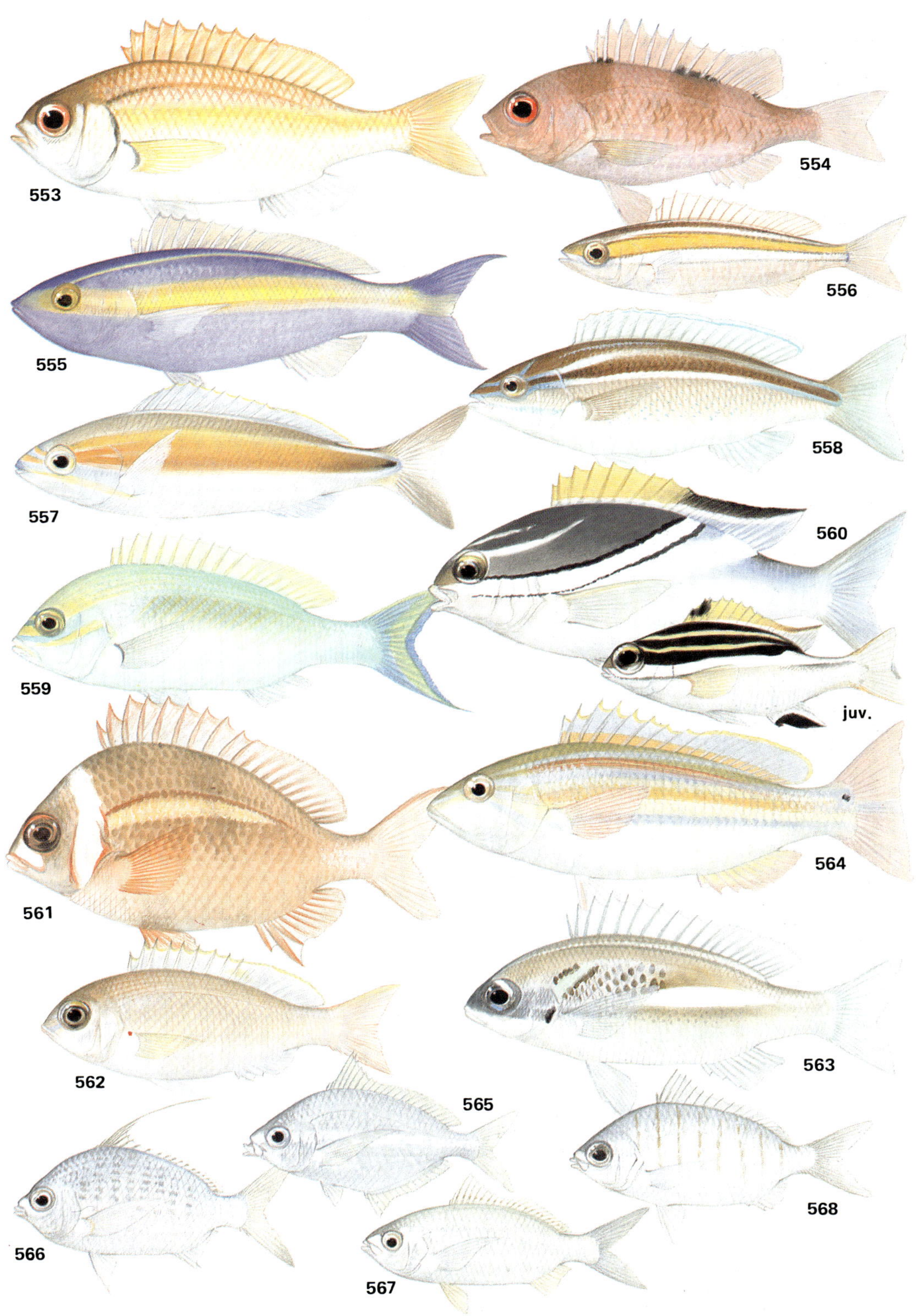

553
554
555
556
557
558
559
560
561
562
563
564
565
566
567
568
juv.

569 GOLD-SADDLED GOATFISH
Parupeneus cyclostomus (Lacepède)
Inhabits coral reefs; has 2 colour varieties: one is entirely yellow, the other is purplish-pink or bluish with gold saddle on top of tail base; Ningaloo Reef northwards; Indo-C. Pacific; to 50 cm; 1.25 kg. ★★★

570 SWARTHY-HEADED GOATFISH
Parupeneus barberinoides (Bleeker)
Inhabits weed-sand areas near coral reefs; distinguished by dark anterior half and pale rear portion with black spot in front of tail base; Ningaloo Reef northwards; mainly W. Pacific; to 25 cm. ★★★

571 DOUBLEBAR GOATFISH
Parupeneus bifasciatus (Lacepède)
Inhabits coral reefs usually over sand or rubble; distinguished by 2 dark saddles on upper half of body; Ningaloo Reef northwards; Indo-C. Pacific; to 35 cm. ★★★

572 YELLOW STRIPED GOATFISH
Parupeneus chrysopleuron (Schlegel)
Inhabits sandy bottoms near rocky areas and coral reefs; distinguished by reddish colour and yellow stripe on upper side, 580 is similar but has patches of teeth on roof of mouth; Geographe Bay northwards; W. Pacific and E. Indian Ocean; to 33 cm; .538 kg. ★★★

573 BLACKSPOT GOATFISH
Parupeneus signatus (Günther)
Inhabits sand-weed areas adjacent to coral reefs and southern rocky reefs; distinguished by alternating light and dark stripes on head and body, and black spot on upper half of tail base; Geographe Bay northwards; W. Pacific and E. Indian Ocean; to 47 cm; 1.651 kg. ★★★

574 INDIAN GOATFISH
Parupeneus indicus (Shaw)
Inhabits sand-weed areas in the vicinity of coral reefs; distinguished by elongate yellow blotch on middle of upper sides and black spot on tail base; Ningaloo Reef northwards; Indo-C. Pacific; to 40 cm; .371 kg. ★★★

575 SPOTTED GOLDEN GOATFISH
Parupeneus cinnabarinus (Cuvier)
Inhabits sand-weed areas in the vicinity of coral or rocky reefs; distinguished by reddish-pink colour and yellow stripes on side, often has small red to brown spot on upper side above pectoral fin; Rottnest Island northwards; Indo-W. Pacific; to 30 cm. ★★★

576 BANDED GOATFISH
Parupeneus multifasciatus (Bleeker)
Inhabits weed-sand areas in the vicinity of coral reefs; distinguished by broad dark bars on side and large spot or bar on tail base; Ningaloo Reef northwards; Indo-C. Pacific; to 30 cm. ★★★

577 SIDESPOT GOATFISH
Parupeneus pleurostigma (Bennett)
Inhabits sand and rubble areas in the vicinity of coral reefs; distinguished by black spot on middle of upper sides followed by pearly white patch; Ningaloo Reef northwards; Indo-C. Pacific; to 30 cm. ★★★

578 GOLDBAND GOATFISH
Upeneus moluccensis (Bleeker)
Inhabits sandy or weed covered areas; distinguished by yellow stripe from eye to tail and narrow stripes on both dorsal fins and upper lobe of tail; Ningaloo Reef northwards; W. Pacific and E. Indian Ocean; to 20 cm. ★★★

579 SUNRISE GOATFISH
Upeneus sulphureus Cuvier
Inhabits sandy or weed covered areas; distinguished by black tip on first dorsal fin and 2-3 gold stripes on side, similar to 581, but lacks stripes on tail; Dampier Archipelago northwards; W. Pacific and E. Indian Ocean; to 23 cm. ★★★

580 OCHRE-BANDED GOATFISH
Upeneus sundaicus Bleeker
Inhabits sandy or weed covered areas; distinguished by yellow stripe on upper sides, similar to 572, but has patches of teeth on roof of mouth, also rear margin of lower lobe of tail usually dark; Derby northwards; N. Indian Ocean and Indo-Australian Archipelago; to 22 cm. ★★★

581 STRIPED GOATFISH
Upeneus vittatus (Forsskål)
Inhabits sandy or weed covered areas; distinguished by 4 orange-yellow stripes on side, black tip on first dorsal fin, and stripes on tail; Shark Bay northwards; Indo-C. Pacific; to 28 cm. ★★★

582 BAR-TAILED GOATFISH
Upeneus tragula Richardson
Inhabits sandy or weed covered areas; similar to 583, but stripe on side dark (reddish brown to blackish), sides mottled with spots, and 8 spines (versus 7) in first dorsal fin; Rottnest Island northwards; Indo-W. Pacific; to 30 cm. ★★★

583 ASYMMETRICAL GOATFISH
Upeneus asymmetricus Lachner
Inhabits sandy or weed covered areas; similar to 582, but stripe on side yellow, fewer small spots on sides, and 7 spines (versus 8) in first dorsal fin; Port Hedland northwards; mainly Indo-Australian Archipelago; to 30 cm. ★★★

GOATFISHES

The goatfishes (family Mullidae) are found mainly in tropical and subtropical seas, usually in the vicinity of reefs. Worldwide about 50-60 species are known. They are characterised by a relatively elongate body, two widely separated dorsal fins, and the presence of a pair of long chin barbels that are used for detecting food. The barbels are also used by males to attract females during courtship. When they are not being used the barbels are tucked tightly under the chin. Goatfishes feed on small fishes and a variety of mainly sand and weed dwelling organisms including worms, shrimps, crabs, molluscs, and echinoderms. They are frequently seen in large schools and most species are considered to be good eating. The maximum size is about 60 cm, but most species are much smaller. The majority of goatfishes are distributed in the Indo-west Pacific region, but there are also a few representatives in the Atlantic and eastern Pacific Oceans.

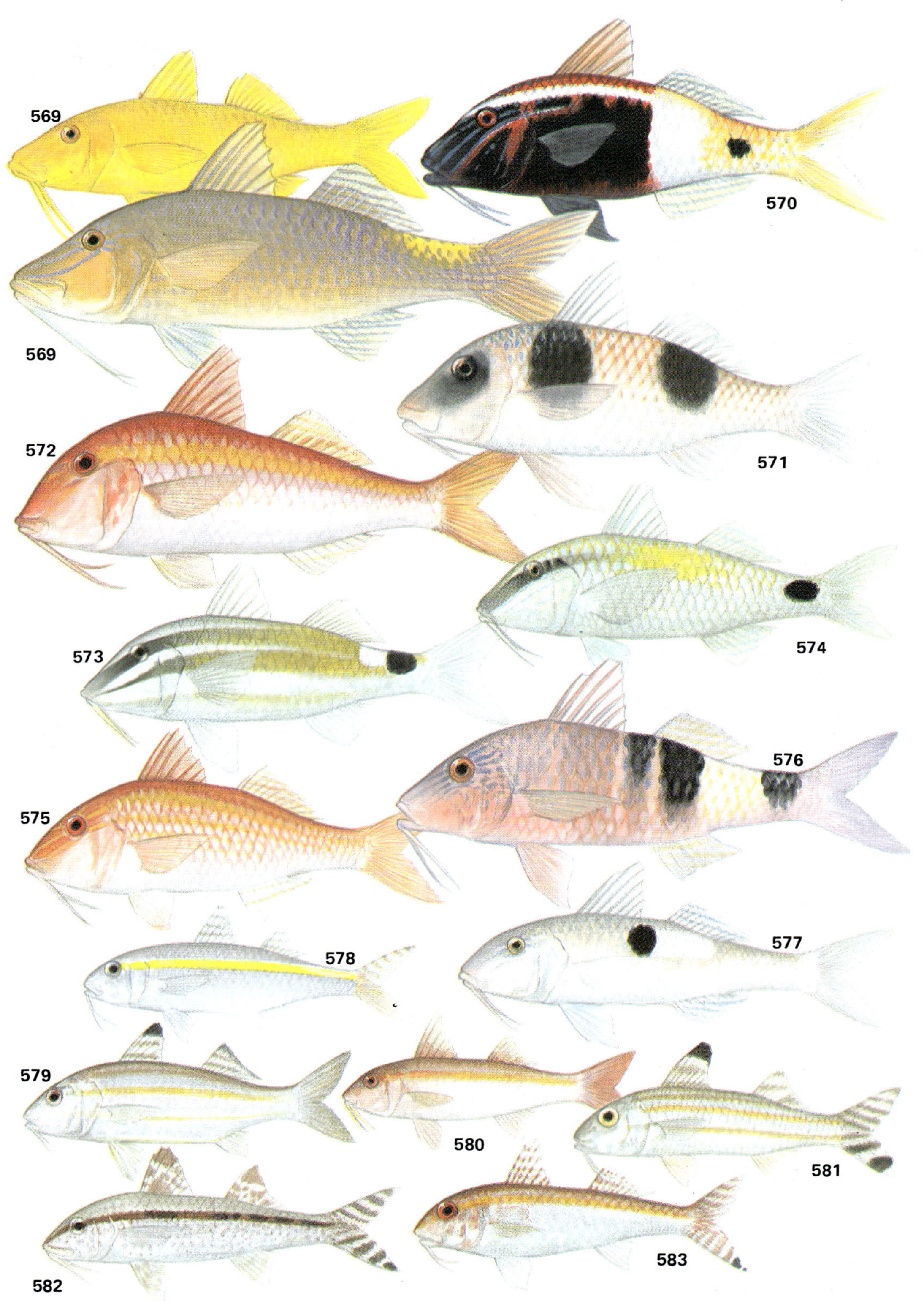

584 BLACK JEW
Protonibea diacanthus (Lacepède)
Inhabits coastal waters, sometimes entering estuaries; distinguished by large size and overall grey to blackish colour, young fish have black spots on the back, dorsal fin, and tail; also known as Spotted croaker or Spotted jewfish; Onslow northwards; Indo-W. Pacific; to 150 cm; 16.2 kg. ★★★

585 ORANGE CROAKER
Argyrosomus sp.
Inhabits coastal waters; distinguished by faint orange tinge on lower third of body when fresh, otherwise without distinctive marks; Exmouth Gulf northwards; Indo-Australian Archipelago; to 35 cm. ★★

586 COITOR CROAKER
Johnius coitor (Hamilton)
Inhabits coastal waters; distinguished by bulging snout with small mouth below, similar to 587, but has dark blotch on gill cover, more soft dorsal rays (30-32 versus 23-26) and more gill rakers on the lower limb of the first gill arch (11 or 12 versus 6-10); Broome northwards; Indo-W. Pacific; to 16 cm. ★★

587 GREEN-BACKED CROAKER
Johnius amblycephalus (Bleeker)
Inhabits coastal waters; distinguished by bulging snout with small mouth below, similar to 586, but distinguished by differences mentioned under that species, also has taller first dorsal fin; Broome northwards; W. Pacific and E. Indian Ocean; to 20 cm. ★★

588 LITTLE JEWFISH
Johnius vogleri (Bleeker)
Inhabits coastal waters including estuaries; distinguished by black outer portion of first dorsal fin and diffuse dark stripe often present on basal part of second dorsal fin; Broome northwards; E. Indian Ocean and Indo-Australian Archipelago; to 22 cm. ★★

589 DIAMOND FISH
Monodactylus argenteus (Linnaeus)
Inhabits estuaries and lower reaches of freshwater streams, also in harbours around wharves and jetties; distinguished by diamond shape and silvery colour; also known as Silver batfish; Dampier northwards; Indo-W. Pacific; to 27 cm; .17 kg.

590 BEACH SALMON
Leptobrama mulleri Steindachner
Inhabits coastal waters, frequently in schools off sandy beaches; distinguished by steel-blue colour on back and a single black-tipped dorsal fin that is set well back on the body; Port Hedland northwards; N. Australia and New Guinea; to 43 cm.

591 NORTHERN SLENDER BULLSEYE
Parapriacanthus unwini (Ogilby)
Inhabits coral reefs, usually in large aggregations in caves; distinguished by semi-transparent appearance, yellowish head, silvery belly and a single dorsal fin; Ningaloo Reef northwards; N. Australia only; to 7.5 cm.

592 BRONZE BULLSEYE
Pempheris analis Waite
Inhabits coral or rocky reefs, usually in caves; similar to 594, but has smaller scales and blackish tips on dorsal and anal fins, and sometimes on lobes of tail; Rottnest Island northwards; Australia only; to 17 cm.

593 OUALAN BULLSEYE
Pempheris oualensis Cuvier
Inhabits coral or rocky reefs, usually in caves; distinguished by relatively large scales and series of dark stripes on side; Rottnest Island northwards; Indo-C. Pacific; to 20 cm.

594 STRIPED BULLSEYE
Pempheris schwenkii Bleeker
Inhabits coral or rocky reefs, usually in caves; similar to 592, but has larger scales, a yellowish tail, blackish anal fin base, and lacks dark fin tips; Rottnest Island northwards; Indo-W. Pacific; to 15 cm.

595 SOUTHERN DRUMMER
Kyphosus gibsoni Ogilby
Inhabits coral reefs, usually in large schools; similar to 597, but lacks well pronounced narrow stripes on side and has 12-13 soft dorsal rays (versus 14 or 15); Point Quobba northwards; Australia only; to 58 cm. ★

596 WESTERN BUFFALO BREAM
Kyphosus cornelii (Whitley)
Inhabits coral and rocky reefs, occurring in large schools; distinguished by dark streak on each tail lobe with white outer margin; Cape Leeuwin to Coral Bay; W.A. only; to 60 cm; 1.2 kg. ★

597 LOW-FINNED DRUMMER
Kyphosus vaigiensis (Quoy & Gaimard)
Inhabits coral reefs, frequently in schools; similar to 595, but has pronounced narrow stripes on side and 14-15 soft dorsal rays (versus 12-13); Abrolhos northwards; Indo-C. Pacific; to 50 cm; 1.43 kg. ★

598 STRIPEY
Microcanthus strigatus (Cuvier)
Inhabits rocky areas and coral reefs, sometimes forms schools around jetties; distinguished by bold pattern of slanting stripes; Cape Leeuwin to Exmouth Gulf; mainly W. and C. Pacific; to 16 cm; .085 kg.

CROAKERS, DRUMMERS, ETC.

Croakers of the family Sciaenidae (584-588) live in a variety of marine and estuarine habitats. They can produce a 'drumming' sound with the aid of their swim bladder. The maximum size is about 2 m, but most are under 50 cm. Some species are important food fishes.

The Diamond Fish, no. 589 (family Monodactylidae) appears to be equally at home in salt or fresh water. It is sometimes found in large aggregations. Young specimens make excellent aquarium pets.

Bullseyes (591-594) of the Indo-Pacific family Pempheridae are cave and crevice dwellers that often occur in large aggregations. They feed mainly on crustaceans, but also consume worms and cephalopods.

Drummers of the family Kyphosidae (595-597) are common inhabitants of reefs and weed beds in tropical and temperate seas. They feed mainly on plants and most are less than 50 cm TL. Although eaten in some localities the flesh is usually not very highly regarded. Some species of *Kyphosus* are reported to produce a mild hallucinogenic effect when consumed.

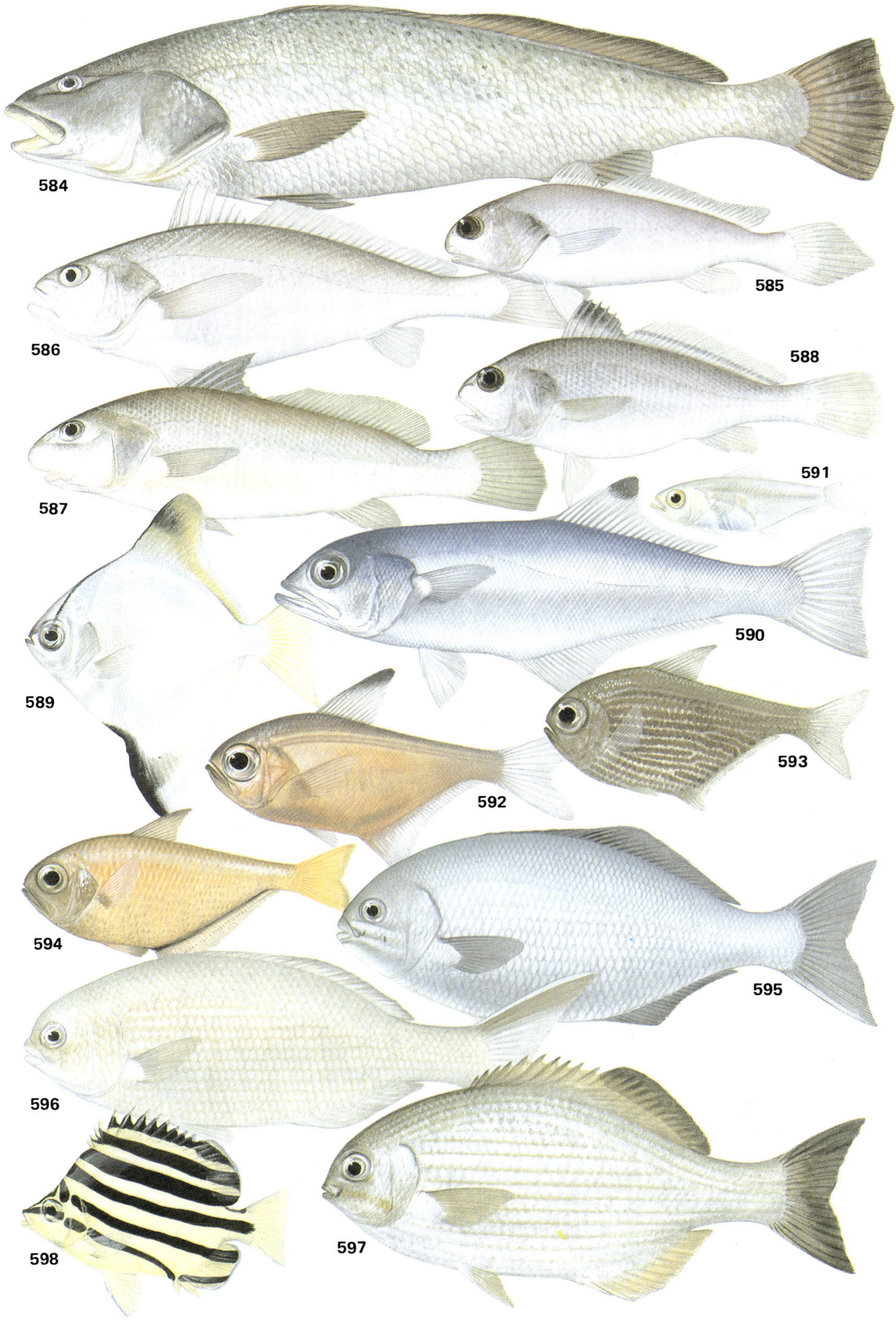

599 SPOTTED ARCHERFISH
Toxotes chatareus (Hamilton)
Inhabits mangrove estuaries and freshwater streams; has ability to knock insects from overhanging vegetation by squirting jets of water from mouth; distinguished from 600, by spotted pattern; Broome northwards; Indo-Australian Archipelago and Andaman Sea; to 30 cm; . 5 kg. ★★★

600 BANDED ARCHERFISH
Toxotes jaculatrix (Pallas)
Inhabits brackish mangrove estuaries; distinguished from 599 by 4 dorsal spines (versus 5) and barred pattern; Derby northwards; Indo-Australian Archipelago and Andaman Sea; to 20 cm. ★★★

601 SICKLEFISH
Drepane punctata (Linnaeus)
Inhabits coastal reefs; distinguished by its triangular shape and vertical rows of small spots; Onslow northwards; Indo-W. Pacific; to 50 cm. ★★★

602 HUMP-HEADED BATFISH
Platax batavianus Cuvier
Inhabits coastal reefs; distinguished by humped forehead and relatively elongate body of adult; Carnarvon northwards; Indo-Australian Archipelago; to 50 cm; 5.795 kg. ★★

603 LONG-FINNED BATFISH
Platax pinnatus (Linnaeus)
Inhabits coral reefs; adult distinguished by slightly protruding snout; Ningaloo Reef northwards; mainly Indo-Australian Archipelago; to 50 cm. ★★

604 ROUND-FACED BATFISH
Platax teira (Forsskål)
Inhabits coral reefs and southern rocky reefs; distinguished by dark blotch below pectoral fin and second blotch above front of anal fin, juvenile with very long dorsal, anal and pelvic fins; Albany northwards; Indo-W. Pacific; to 60 cm. ★★

605 NARROW-BANDED BATFISH
Platax orbicularis (Forsskål)
Inhabits coral reefs and inshore areas; similar to 604 but adults generally have shorter dorsal and anal fins and lack a pronounced blotch below pectoral fin; juvenile effectively mimics dead leaves; Point Quobba northwards; Indo-C. Pacific to 50 cm. ★★★

606 SHORT-FINNED BATFISH
Zabidius novemaculeatus (McCulloch)
Inhabits coastal reefs; similar to 602, but has 9 dorsal spines (versus 7) and lacks pronounced hump on forehead of adults; Onslow northwards; N. Australia only; to 45 cm; 0.800 kg. ★★

607 THREADFIN SCAT
Rhinoprenes pentanemus Munro
Inhabits coastal seas; distinguished by bulbous snout, laterally compressed body and long dorsal and anal fin filaments; northern Kimberley coast. N. Australia and S. New Guinea; to 15 cm. ★★

608 SPOTTED SCAT
Scatophagus argus (Linnaeus)
Text on page 96.

609 STRIPED BUTTERFISH
Selenotoca multifasciata (Richardson)
Text on page 96.

ARCHERS AND BATS

Archerfishes of the family Toxotidae (599-600) exhibit one of nature's most remarkable feeding adaptations. They are renowned for their ability to knock down insects with a squirt of water. Archers produce their aqueous bullets by suddenly pulling in their gill flaps, forcing water through a tube formed by a groove on the roof of the mouth and the tongue. Their accuracy is amazing, considering the fish must compensate for the angle of refraction of light. Because of the bending of light rays as they enter the water, archerfish view their targets in positions different from the true ones. Nevertheless, their accuracy is uncanny. They are reputed to be able to hit a small insect up to a distance of 3 metres!

The Spadefishes of the family Ephippididae (601-606) are distributed throughout tropical and subtropical seas. Most Indo-Pacific species belong to the genus *Platax* and are commonly known as batfishes. They are characterised by a very deep, nearly circular body and often have distinctively elongate dorsal and anal fins. Young batfish have particularly long, flowing fins and may possess highly contrasted patterns with vertical bars. They are widely prized as aquarium specimens. The largest species grow to about 50 cm TL. Batfishes and other ephippidids are often encountered around coral reefs either individually or sometimes in large schools. The diet includes small benthic invertebrates and also planktonic items. Young batfishes sometimes mimic dead leaves by swimming on their sides.

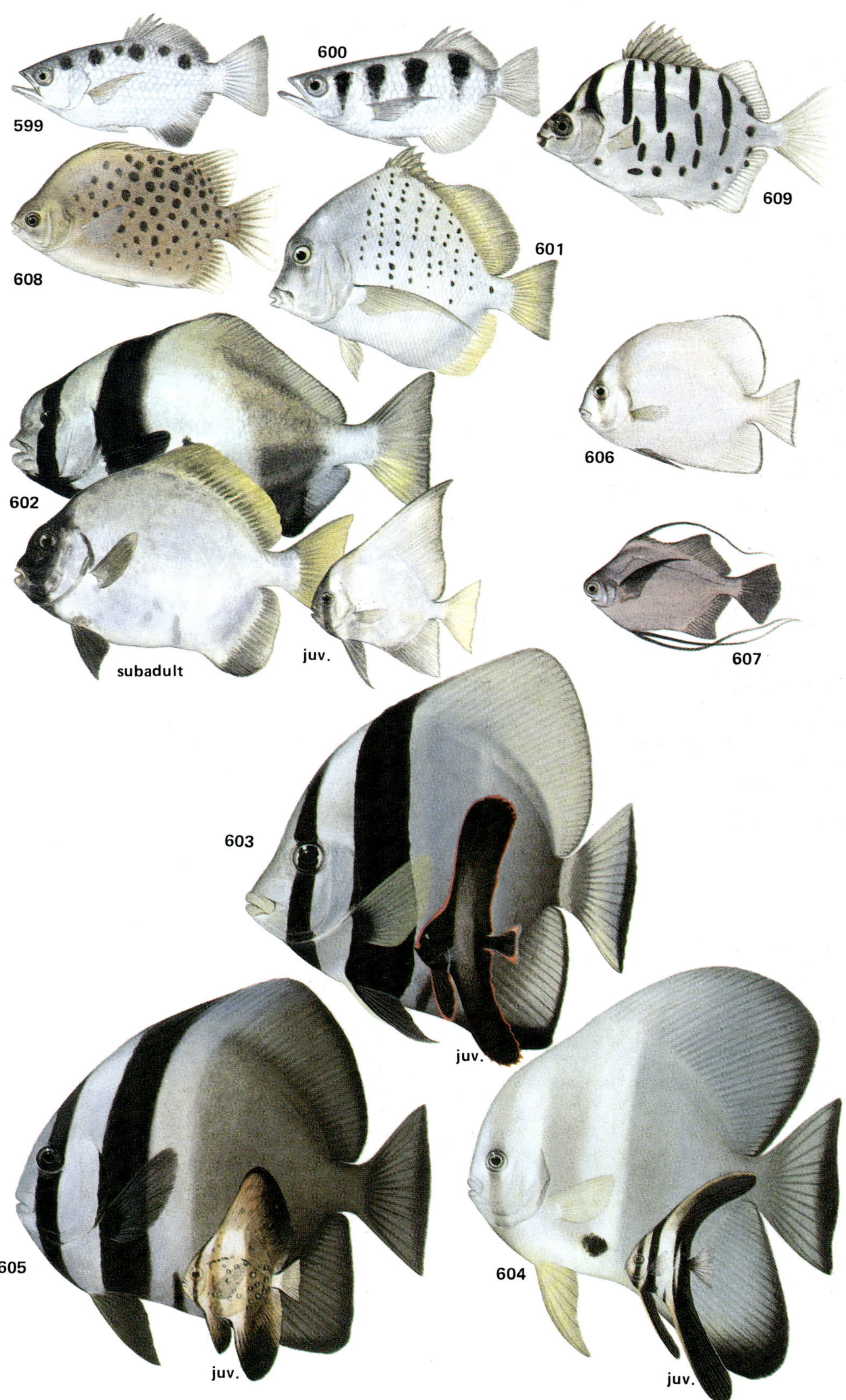

608 SPOTTED SCAT
Scatophagus argus (Linnaeus)
Inhabits brackish mangrove estuaries and fresh-water streams; similar to 608, but has numerous round spots on side; Dampier northwards; Indo-W. Pacific; to 33 cm. ★★

609 STRIPED BUTTERFISH
Selenotoca multifasciata (Richardson)
Inhabits brackish mangrove estuaries and freshwater streams; similar to 609 but has more silvery body colour and combinations of bars and spots; also known as Spotband Scat; Shark Bay northwards; Indo-Australian Archipelago; to 28 cm. ★★

610 PHILIPPINE BUTTERFLYFISH
Chaetodon adiergastos Seale
Inhabits coral reefs; distinguished by broad eye bar and narrow diagonal stripes on side; Dampier Archipelago; Indo-Australian Archipelago; to 15 cm.

611 WESTERN BUTTERFLYFISH
Chaetodon assarius Waite
Inhabits rocky reefs and sandy-weed flats; distinguished by several rows of small dots on sides; Recherche Archipelago to Monte Bello Islands; W.A. only; to 16 cm.

612 THREADFIN BUTTERFLYFISH
Chaetodon auriga Forsskål
Inhabits coral reefs and sandy-weed flats; distinguished by chevron markings and yellow posterior; Rottnest Island northwards; Indo-C. Pacific; to 23 cm.

613 GOLDENSTRIPED BUTTERFLYFISH
Chaetodon aureofasciatus Macleay
Inhabits coral reefs; distinguished by overall yellow-orange colour and pair of orange stripes on head and anterior part of body; Coral Bay northwards; Australia and New Guinea; to 14 cm.

614 SPECKLED BUTTERFLYFISH
Chaetodon citrinellus Cuvier
Inhabits shallow coral reefs, where live coral is sparse; distinguished by small dots on yellow background; Rottnest Island northwards; Indo-C. Pacific; to 13 cm.

615 SADDLED BUTTERFLYFISH
Chaetodon ephippium Cuvier
Inhabits coral reefs; distinguished by large black saddle; Coral Bay northwards; Indo-C. Pacific; to 23 cm.

616 LINED BUTTERFLYFISH
Chaetodon lineolatus Cuvier
Inhabits coral reefs; distinguished by narrow black stripes on sides and broad black area along dorsal fin base; Rottnest Island northwards; Indo-C. Pacific; to 31 cm.

617 MEYER'S BUTTERFLYFISH
Chaetodon meyeri Bloch & Schneider
Inhabits coral reefs; distinguished by black diagonal stripes; Ningaloo Reef northwards; Indo-W. Pacific; to 20 cm.

618 KLEIN'S BUTTERFLYFISH
Chaetodon kleinii Bloch
Inhabits coral reefs; distinguished by combination of bars on front of body and golden brown area on rear half; Coral Bay northwards but rarely seen; Indo-C. Pacific; to 13 cm.

619 BLUESPOT BUTTERFLYFISH
Chaetodon plebeius Cuvier
Inhabits coral reefs; distinguished by ovate blue spot on yellow background; Rottnest Island northwards; Indo-W. Pacific; to 13 cm.

620 RACOON BUTTERFLYFISH
Chaetodon lunula Lacepède)
Inhabits coral reefs; distinguished by black "mask" and diagonal black bar behind head with golden colour on lower sides; Rottnest Island northwards; Indo-C. Pacific; to 21 cm.

621 ORNATE BUTTERFLYFISH
Chaetodon ornatus Cuvier
Inhabits coral reefs; distinguished by brown-orange diagonal stripes; Ningaloo Reef northwards; mainly W. and C. Pacific; to 18 cm.

COLOURFUL BUTTERFLYFISHES

Butterflyfishes of the family Chaetodontidae (Plates 39-40) are renowned for their striking colour patterns, delicate shapes, and graceful swimming movements. The family contains 116 species which occur mainly in tropical seas around coral reefs. Most of the species dwell in depths of less than 20 m, but some are restricted to deeper sections of the reef, to at least 200 m. Butterflyfishes are active during daylight hours and seek shelter close to the reef's surface during the night. They often assume a drab nocturnal colour pattern. Most species are restricted to a relatively small area of the reef, perhaps an isolated patch reef or part of a more extensive reef system. They travel extensively throughout their home range foraging for food. Many species feed on live coral polyps, others consume a mixed diet consisting of small benthic invertebrates and algae. A few species, for example *Heniochus diphreutes* (633), feed in midwater on zooplankton. Young butterflyfishes are highly prized as aquarium fishes. Most species grow to a maximum length that is under 30 cm.

Butterflyfishes are one of the most conspicuous inhabitants of tropical reefs and have attracted the attention of behavioural scientists. Some species are generally found solitarily or occasionally in small groups. Others form pairs and depending on the frequency of their pairing tendencies are termed weakly or strongly paired. The weak category includes such species as 611, 614, 620 and 621. Examples of strong pairing species are 615, 622, 623, 626, and 627. There is strong evidence that fishes in the latter category form lifetime relationships.

Scats (608-609) in the family Scatophagidae are closely related. They generally occur around wharves in harbours, mangrove estuaries, and the lower reaches of streams. Juveniles are popular aquarium fish.

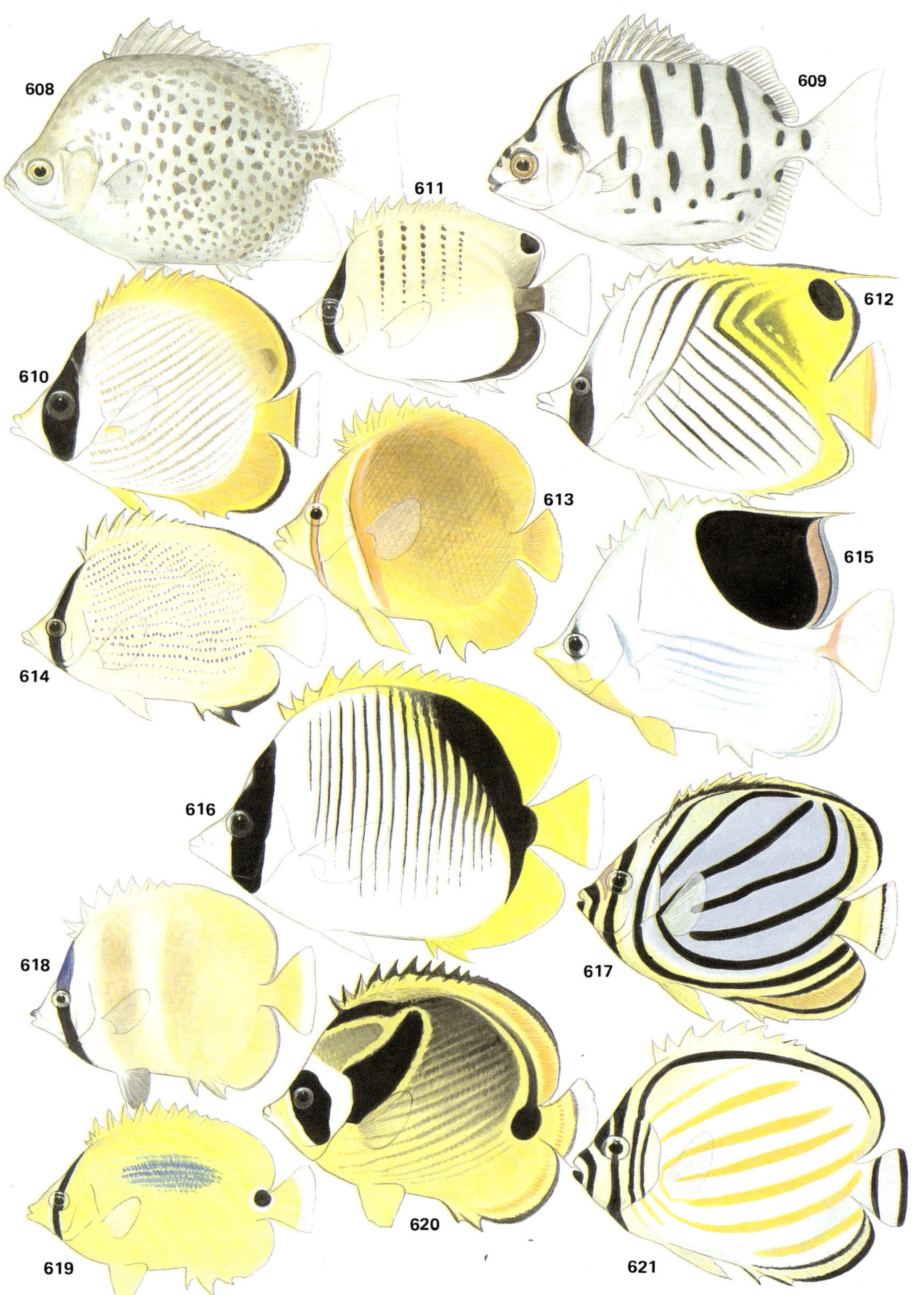

622 TEARDROP BUTTERFLYFISH
Chaetodon unimaculatus Bloch
Inhabits coral reefs; distinguished by black spot on white background and yellow fins; Ningaloo Reef northwards; Indo-C. Pacific; to 23 cm.

623 SPOTBANDED BUTTERFLYFISH
Chaetodon punctatofasciatus Cuvier
Inhabits coral reefs; distinguished by bars on back and spots below; Ningaloo Reef northwards; mainly W. Pacific; to 13 cm.

624 OVALSPOT BUTTERFLYFISH
Chaetodon speculum Cuvier
Inhabits coral reefs; distinguished by round spot on yellow background; Abrolhos northwards; mainly W. Pacific; to 15 cm.

625 DOUBLESADDLE BUTTERFLYFISH
Chaetodon ulietensis Cuvier
Inhabits coral reefs; distinguished by 2 broad blackish bars and series of narrow vertical lines on white background; Ningaloo Reef northwards; mainly W. Pacific; to 14 cm.

626 CHEVRONED BUTTERFLYFISH
Chaetodon trifascialis Quoy & Gaimard
Inhabits coral reefs; distinguished by somewhat triangular-shaped body and narrow chevron markings; Ningaloo Reef northwards; Indo-C. Pacific; to 18 cm.

627 REDFIN BUTTERFLYFISH
Chaetodon trifasciatus Park
Inhabits coral reefs; distinguished by dark stripes on orange to purple background and reddish anal fin; Abrolhos northwards; Indo-C. Pacific; to 15 cm.

628 ORANGEBANDED CORALFISH
Coradion chrysozonus (Cuvier)
Inhabits coral reefs and trawl grounds; distinguished by orange-brown bars and black pelvic fins; Shark Bay northwards; mainly Indo-Australian Archipelago; to 13 cm.

629 MARGINED CORALFISH
Chelmon marginalis Richardson
Inhabits coral reefs; distinguished by long snout and orange bars; Shark Bay northwards; N. Australia only; to 20 cm.

630 LONGNOSED BUTTERFLYFISH
Forcipiger flavissimus Jordan & McGregor
Inhabits coral reefs; distinguished by long snout, black colour on upper half of head and yellow colour of body; Ningaloo Reef northwards; Indo-E. Pacific; to 17 cm.

631 PENNANT BANNERFISH
Heniochus chrysostomus Cuvier
Inhabits coral reefs; distinguished by short "banner", and broad black band across head that is continuous with pelvic fins; North West Shelf reefs; mainly W. and C. Pacific; to 17 cm.

632 SINGULAR BANNERFISH
Heniochus singularius Smith & Radcliffe
Inhabits coral reefs; distinguished by short "banner", slight hump on forehead, and yellow dorsal and caudal fins; Ningaloo Reef, northwards; mainly W. and C. Pacific; to 25 cm.

633 SCHOOLING BANNERFISH
Henichus diphreutes Jordan
Inhabits sandy areas around outcrops; usually in schools; distinguished by elongated dorsal rays, resembles 634, but has longer snout and rounded anal fin; Trimouille I. northwards; Indo-C. Pacific; to 20 cm.

634 LONGFIN BANNERFISH
Heniochus acuminatus (Linnaeus)?
Inhabits coral reefs, often in pairs or alone; distinguished by elongated dorsal rays, resembles 633, but shorter snout and angular anal fin; Cervantes northwards; Indo-C. Pacific; to 20 cm.

635 HUMPHEAD BANNERFISH
Heniochus varius (Cuvier)
Inhabits coral reefs; distinguished by lack of "banner" and forehead bump and "horns"; North West Shelf reefs; mainly W. and C. Pacific; to 18 cm.

636 OCELLATE CORALFISH
Parachaetodon ocellatus (Cuvier)
Inhabits coral reefs; distinguished by tall triangular dorsal fin and orange-brown bars; Abrolhos northwards; Indo-W. Pacific; to 18 cm.

TWIN SPECIES

Many reef fishes have close relatives or 'sister' species that are very similar in overall appearance. These pairs have obviously evolved from a common ancestral population that became fragmented by such barriers as shifting ocean currents, sea temperature changes, and emergent land resulting from lowered sea levels or tectonic processes. In cases where the ancestral stock is widely distributed it may become fragmented into more than two populations evolving into 'complexes' of several closely allied species. As the barriers may be only temporary, even though persisting for thousands of years (a short span in terms of geological time), members of a pair may once again occur in the same area.

The two similar species of *Heniochus*, 633 and 634, are good examples of this process. Although they now have overlapping distributions they are more or less separated ecologically. The Schooling Bannerfish (633) forms groups which feed on plankton high above the bottom whereas the Longfin Bannerfish (634) lives alone or in small groups which forage on bottom dwelling invertebrates. Other species-pairs continue to exist more or less separately except for a relatively small overlap zone. For example, a number of pairs have one member in the Indian Ocean and another in the Pacific. In most cases the species found in W.A. seas is the Pacific form rather than the expected Indian Ocean species.

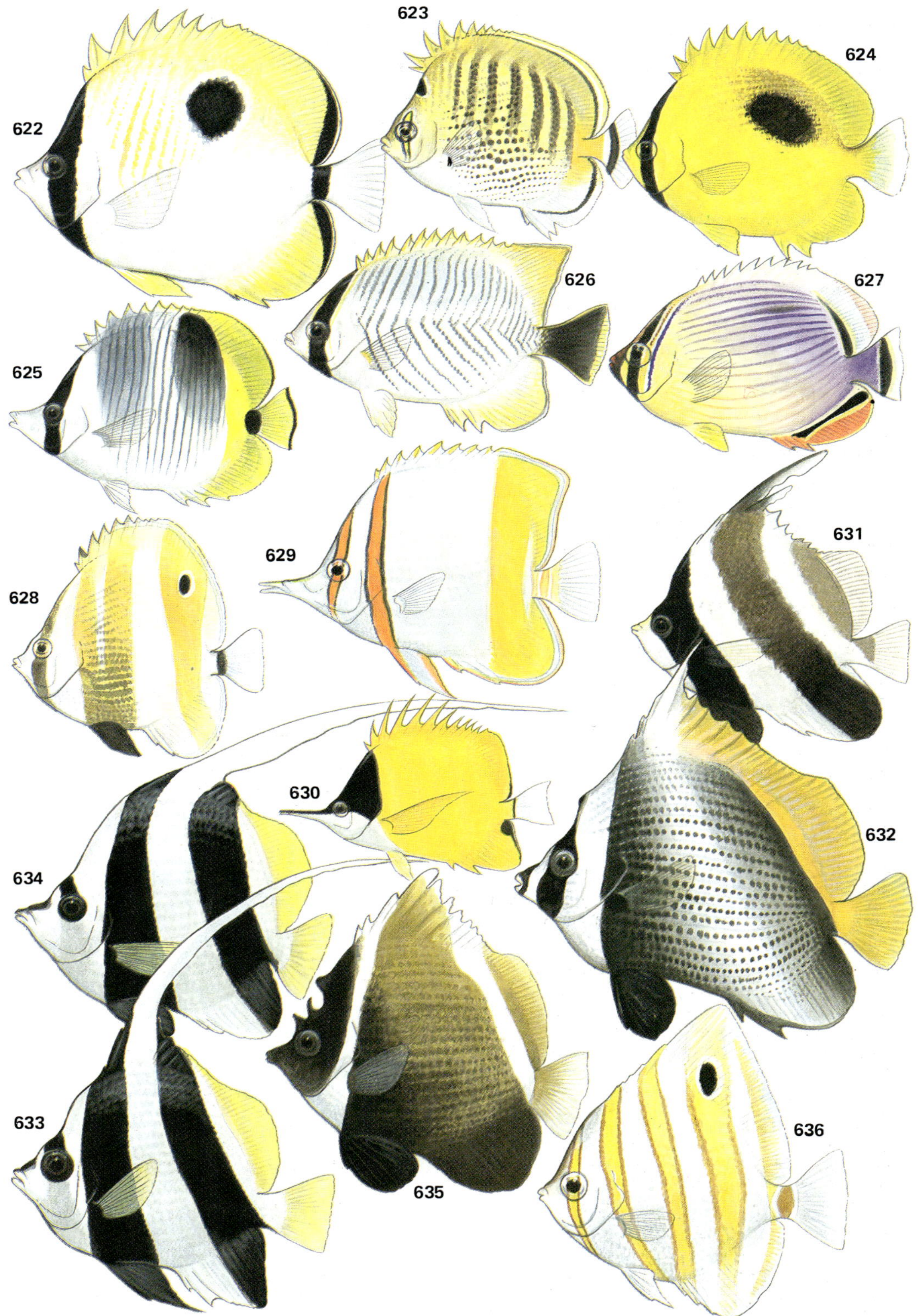

622
623
624
625
626
627
628
629
631
630
632
634
633
635
636

637 THREE-SPOT ANGELFISH
Apolemichthys trimaculatus (Lacepède)
Inhabits offshore coral reefs, usually below 15 m depth; distinguished by blue lips, black spot on forehead, and broad black margin on anal fin; Ningaloo Reef northwards; Indo-W. Pacific; to 20 cm. ★★

638 BICOLOR ANGELFISH
Centropyge bicolor (Bloch)
Inhabits coral reefs, but very rare on our coast; distinguished by blue band through eye, and yellow-blue combination on head and body; Point Quobba northwards; mainly central and W. Pacific; to 15 cm.

639 EIBL'S ANGELFISH
Centropyge eibli Klausewitz
Inhabits coral reefs, but rare on our coast, has only been seen on Ningaloo Reef complex; distinguished by black tail and narrow brown or orange bars on side; central and E. Indian Ocean; to 11 cm.

640 KEYHOLE ANGELFISH
Centropyge tibicen (Cuvier)
Inhabits coral reefs; distinguished by vertically elongate white spot on middle of side and yellow margin on pelvic and anal fins; Abrolhos northwards; mainly W. Pacific; to 14 cm.

641 SCRIBBLED ANGELFISH
Chaetodontoplus duboulayi (Günther)
Inhabits rubble, soft bottoms, or open flat bottom areas with rock, coral, sponge, and seawhip outcrops; distinguished by dark bar through eye, yellow arc along back, and yellow tail, female (not shown) is dark on sides or has numerous blue or yellowish spots, male has broken wavy blue lines; Shark Bay northwards; N. Australia and New Guinea; to 28 cm. ★★

642 YELLOWTAIL ANGELFISH
Chaetodontoplus personifer McCulloch
Inhabits similar areas to 641, usually in the vicinity of reefs; distinguished by blue "mask" with yellow spots on cheek and above eye, juvenile is black with broad white bar behind eye, and has yellow pelvic fins and tail; Abrolhos northwards; N. Australia and Taiwan; to 35 cm. ★★

643 SIX-BANDED ANGELFISH
Pomacanthus sexstriatus (Cuvier)
Inhabits coral reefs; distinguished by white bar behind eye, 4-5 dark bars on side, and numerous blue spots on body and fins, juvenile similar to 644, but white bars generally broader; Ningaloo Reef northwards; Indo-Australian Archipelago; to 45 cm. ★★

644 BLUE-GIRDLED ANGELFISH
Pomacanthus navarchus (Cuvier)
Inhabits protected lagoon coral reefs; distinguished by broad yellow-orange area on sides with small dark spots, yellow-orange dorsal and tail fins, and bright blue stripe from snout to belly, also narrow blue margins on most fins; offshore reefs of North West Shelf; Indo-Australian Archipelago; to 28 cm. ★★

645 EMPEROR ANGELFISH
Pomacanthus imperator (Bloch)
Inhabits coral reefs; distinguished by blue-edged black band through eye and narrow yellow stripes on side, juveniles dark with narrow white lines, forming concentric circles on rear part of body; Shark Bay northwards; Indo-C. Pacific; to 31 cm. ★★

646 BLUE ANGELFISH
Pomacanthus semicirculatus (Cuvier)
Inhabits coral reefs; distinguished by brown to yellowish colour with overall bluish hue imparted by numerous blue spots on body and fins, also has blue margins on cheek, gill cover, and most fins, juveniles similar to 643 and 644, but posterior white bars are strongly curved; Abrolhos northwards; Indo-W. Pacific; to 38 cm. ★★

647 REGAL ANGELFISH
Pygoplites diacanthus (Boddaert)
Inhabits coral reefs; distinguished by alternating dark-edged white and yellow-orange cross bars; offshore reefs of North West Shelf; Indo-C. Pacific; to 25 cm.

ANGELFISHES

Angelfishes of the family Pomacanthidae are close relatives of the butterflyfishes (Plates 39-40) and until recently were considered to belong in the same family. Worldwide there are 76 known species. Most are inhabitants of tropical seas, being found mainly in the vicinity of coral reefs. They occur both as solitary individuals or in aggregations. Many species inhabit shallow water, from only 2-3 m down to 10 to 15 m depth. Others are restricted to deep water (to at least 75 m depth). Angelfishes are favourite aquarium pets, well known for their brilliant array of colour patterns. Many species exhibit dramatic changes from the juvenile to adult stage. Most angelfishes are dependent on the presence of shelter in the form of boulders, caves, and coral crevices. Typically, most species are somewhat territorial and spend daylight hours near the bottom in search of food. The diet varies according to species; some feed almost exclusively on algae, others prefer mainly sponges supplemented by a variety of benthic invertebrates, and a few are midwater zooplankton feeders. Divers are sometimes startled by the powerful drumming or thumping sound which is produced by large adult angels in the genus *Pomacanthus* (643-646).

637
638
639
640
641
642
643
juv.
644
645
juv.
646
juv.
647
juv.

648 BANDED SERGEANT
Abudefduf septemfasciatus (Cuvier)
Inhabits shallow, wave-swept reefs; distinguished by 7 grey bars; Shark Bay northwards; Indo-C. Pacific; to 22 cm.

649 BLACKSPOT SERGEANT MAJOR
Abudefduf sordidus Forsskål
Inhabits shallow, wave-swept reefs; similar to 648 but has black spot on top of tail base; Abrolhos northwards; Indo-C. Pacific; to 22 cm; .44 kg.

650 SERGEANT MAJOR
Abudefduf vaigiensis (Quoy & Gaimard)
Inhabits coral, rocky, and weedy reefs; distinguished by 5 dark bars; Rottnest Island northwards; Indo-C. Pacific; to 22 cm; .299 kg.

651 NARROW-BANDED SERGEANT MAJOR
Abudefduf bengalensis (Bloch)
Inhabits coral and weedy reefs; distinguished from other *Abudefduf* by narrower bars and rounded lobes of tail; also known as Bengal Sergeant; Dongara northwards; Indo-W. Pacific; to 18 cm; .155 kg.

652 SCISSORTAIL SERGEANT
Abudefduf sexfasciatus (Lacepéde)
Inhabits coral and weedy reefs; similar to 650, but has black streaks on tail fin; Rottnest Island northwards; Indo-C. Pacific; to 22 cm; .104 kg.

653 STAGHORN DAMSEL
Amblyglyphidodon curacao (Bloch)
Inhabits coral reefs, often among "staghorn" corals; distinguished by ovate shape and faint bars; Coral Bay northwards; mainly W. Pacific; to 13 cm.

654 BANDED DAMSEL
Dischistodus darwiniensis (Whitley)
Inhabits coral and rocky reef near sand; distinguished by brown bands on side and 2 dark spots on dorsal fin; Dampier northwards; Indo-Australian Archipelago; to 15 cm.

655 HONEYHEAD DAMSEL
Dischistodus prosopotaenia (Bleeker)
Inhabits coral reefs near sand; distinguished by pale band through mid body; Coral Bay northwards; mainly W. Pacific; to 20 cm.

656 REGAL DEMOISELLE
Neopomacentrus cyanomos (Bleeker)
Inhabits coral reefs; distinguished by slender shape and white on rear of dorsal, anal, and tail fins; Point Quobba northwards; Indo-W. Pacific; to 10 cm.

657 BROWN DEMOISELLE
Neopomacentrus filamentosus (Macleay)
Inhabits coral reefs; distinguished by slender shape, blue border on fins and dark margins on tail; Shark Bay northwards; Indo-Australian Archipelago; to 9 cm.

658 YELLOWTAIL DEMOISELLE
Neopomacentrus azysron (Bleeker)
Inhabits coral reefs; distinguished by slender shape, dark "ear" spot and yellow tail; Shark Bay northwards; Indo-W. Pacific; to 9 cm.

659 LAGOON DAMSEL
Hemiglyphidodon plagiometopon (Bleeker)
Inhabits coral reefs; distinguished by plain brown colouration, pointed snout and lack of spines along the margin of the cheek; North West Shelf reefs; mainly W. Pacific; to 22 cm.

660 BLACK DAMSEL
Paraglyphidodon melas (Cuvier)
Inhabits coral reefs, often with soft corals; adults entirely black; Coral Bay northwards; Indo-W. Pacific; to 16 cm.

661 BEHN'S DAMSEL
Paraglyphidodon nigroris (Cuvier)
Inhabits coral reefs; adults mainly dark brown, juveniles brilliant yellow with pair of black stripes; Coral Bay northwards; mainly W. Pacific; to 14 cm.

DAMSELS OF THE SEA

The Damselfishes of the family Pomacentridae (Plates 42-44) are one of the most abundant groups of coral reef fishes. Approximately 320 species occur worldwide including about 120 from Australian seas. Most inhabit the tropics, but a number of species live in cooler temperate waters. They display remarkable diversity with regards to habitat preference, feeding habits, and behaviour. Coloration is highly variable ranging from drab hues of brown, grey and black to brilliant combinations of orange, yellow, and neon blue. A number of species have juvenile stages characterised by a yellow body with bright blue stripes crossing the upper head and back. Most damselfishes are territorial, particularly algal feeding species such as 689 and 691 on Plate 44. They zealously defend their small plot against all intruders regardless of size. Damsels exhibit a highly stereotyped mode of reproduction in which one or both partners clear a next site on the bottom and engage in courtship displays of rapid swimming and fin extension. Males generally guard the eggs which are attached to the bottom by adhesive strands. The eggs hatch within about 2-7 days and the fragile larvae rise to the surface. They are transported by ocean currents for periods which vary between about 10-50 days, depending on the species. Eventually the young fish settle to the bottom and their largely transparent bodies quickly assume the juvenile coloration. The growth rate of juveniles generally ranges from about 5-15 mm per month and gradually tapers off as maturity approaches. There is very little reliable data concerning their longevity, but it appears they are capable of living to at least an age of 10 years. An unusual type of sex change has been documented in the anemonefishes (Plate 43; 662-665) in which males eventually become females. Damselfishes feed on a wide variety of plant and animal material. Generally, the drab-coloured species feed mainly on algae, whereas many of the brightly patterned species and also members of the genus Chromis (Plate 43; 668-673) obtain their nourishment from current-borne plankton.

648
649
650
651
652
653
654
655
656
657
658
659
660
661
juv.
juv.
juv.

662 CLARK'S ANEMONEFISH
Amphiprion clarkii Cuvier
Inhabits coral reefs; associated with large sea anemones; distinguished by 2 white bars and pale tail; Abrolhos northwards; Indo-W. Pacific; to 13 cm.

663 FALSE CLOWN ANEMONEFISH
Amphiprion ocellaris Cuvier
Inhabits coral reefs in protected waters; associated with large sea anemones; distinguished by 3 white bars on head and body and black submarginal bands on fins; Broome northwards; mainly W. Pacific and Andaman Sea; to 7.5 cm.

664 PINK ANEMONEFISH
Amphiprion perideraion Bleeker
Inhabits coral reefs; associated with large sea anemones; distinguished by a single narrow bar on each side of head; Ningaloo Reef northwards; mainly W. and C. Pacific; to 10 cm.

665 ORANGE ANEMONEFISH
Amphiprion sandaracinos Allen
Inhabits coral reefs; associated with large sea anemones; distinguished by a white stripe along the base of the dorsal fin; Ningaloo Reef northwards; mainly W. Pacific; to 13 cm.

666 RED ANEMONEFISH
Amphiprion rubrocinctus Richardson
Inhabits coral reefs; associated with large sea anemones; distinguished by black to reddish colour and white bar on head (sometimes weakly developed or absent); Point Quobba northwards; N.W. Australia only; to 13 cm.

667 BIG-LIP DAMSEL
Cheiloprion labiatus Day
Inhabits coral reefs; usually amongst beds of branching *Acropora* corals; feeds partly on live corals; distinguished by enlarged lips; juveniles are primarily blackish with a neon-blue stripe along the upper back. Dampier Archipelago northwards; Indo-W. Pacific; to 8 cm.

668 BLACK-AXIL CHROMIS
Chromis atripectoralis Welander & Schultz
Inhabits coral reefs; forms large aggregations above coral bommies; distinguished from 669 by the black "arm pit" of the pectoral fin; Abrolhos northwards; Indo-C. Pacific; to 10 cm.

669 BLUE-GREEN CHROMIS
Chromis viridis (Cuvier)
Inhabits coral reefs; forms aggregations that shelter in branching coral; distinguished from 668 by smaller size and lack of black colour on inside of pectoral fin base; Point Quobba northwards; Indo-C. Pacific; to 8 cm.

670 GREEN CHROMIS
Chromis cinerascens (Cuvier)
Inhabits coral reefs; distinguished by overall dusky green appearance and faint longitudinal lines along the side of the body; Dampier Archipelago northwards; W. Pacific and E. Indian Ocean; to 10 cm.

671 SMOKEY CHROMIS
Chromis fumea (Tanaka)
Inhabits coral reefs, usually in deeper water (10-15 m+); distinguished by dark caudal fin margins and pearly spot behind dorsal fin; Point Quobba northwards; mainly W. Pacific; to 9 cm.

672 BICOLOR CHROMIS
Chromis margaritifer Fowler
Inhabits coral reefs; distinguished by two-tone colour pattern; Point Quobba northwards; mainly W. Pacific; to 7.5 cm.

673 WEBER'S CHROMIS
Chromis weberi Fowler & Bean
Inhabits coral reefs; distinguished by dark scale edges forming a network pattern and by the dark tips on the tail; Point Quobba northwards; Indo-C. Pacific; to 12 cm.

674 THREE-SPOT DASCYLLUS
Dascyllus trimaculatus (Rüppell)
Inhabits coral reefs, juveniles sometimes associated with branching corals or large sea anemones; distinguished by white spot on forehead and similar spot on each side of body, these gradually fade with increased size; Abrolhos northwards; Indo-C. Pacific; to 13 cm.

675 RETICULATED DASCYLLUS
Dascyllus reticulatus (Richardson)
Inhabits coral reefs, occurs in aggregations around small coral formations; distinguished by black bar on front of body; Abrolhos northwards; Indo-C. Pacific; to 9 cm.

676 HUMBUG DASCYLLUS
Dascyllus aruanus (Linneaus)
Inhabits coral reefs, usually seen in aggregations around small coral formations; distinguished by 3 black bars; Abrolhos northwards; Indo-C. Pacific; to 8 cm.

SYMBIOTIC DAMSELS

Symbiosis is a biological term that literally means 'living together'. There are many examples of this phenomenon in nature, but one of the most colourful and well documented involves damselfishes belonging to the genus *Amphiprion* (662-665). The genus contains 26 species which occur throughout the tropical Indo-west Pacific region. All of the species are associated with large sea anemones that normally sting fishes which make contact with their tentacles. However, anemonefishes are never stung and live amongst the tentacles with impunity. There are two factors which contribute to the fishes 'immunity'. Apparently the distinctive swimming behaviour of the fish and special chemicals in the external mucus coat prevent the anemone from firing its nematocysts (stinging cells). Anemonefishes gain security from predators because of their association and the anemones themselves are protected from coelenterate feeders such as butterflyfishes. The fishes also keep their host anemones free of debris and experiments indicate that the fishes directly contribute to their 'health', keeping the tentacles in a robust condition. Juveniles of the Three-spot Dascyllus (674) sometimes are associated with anemones as well.

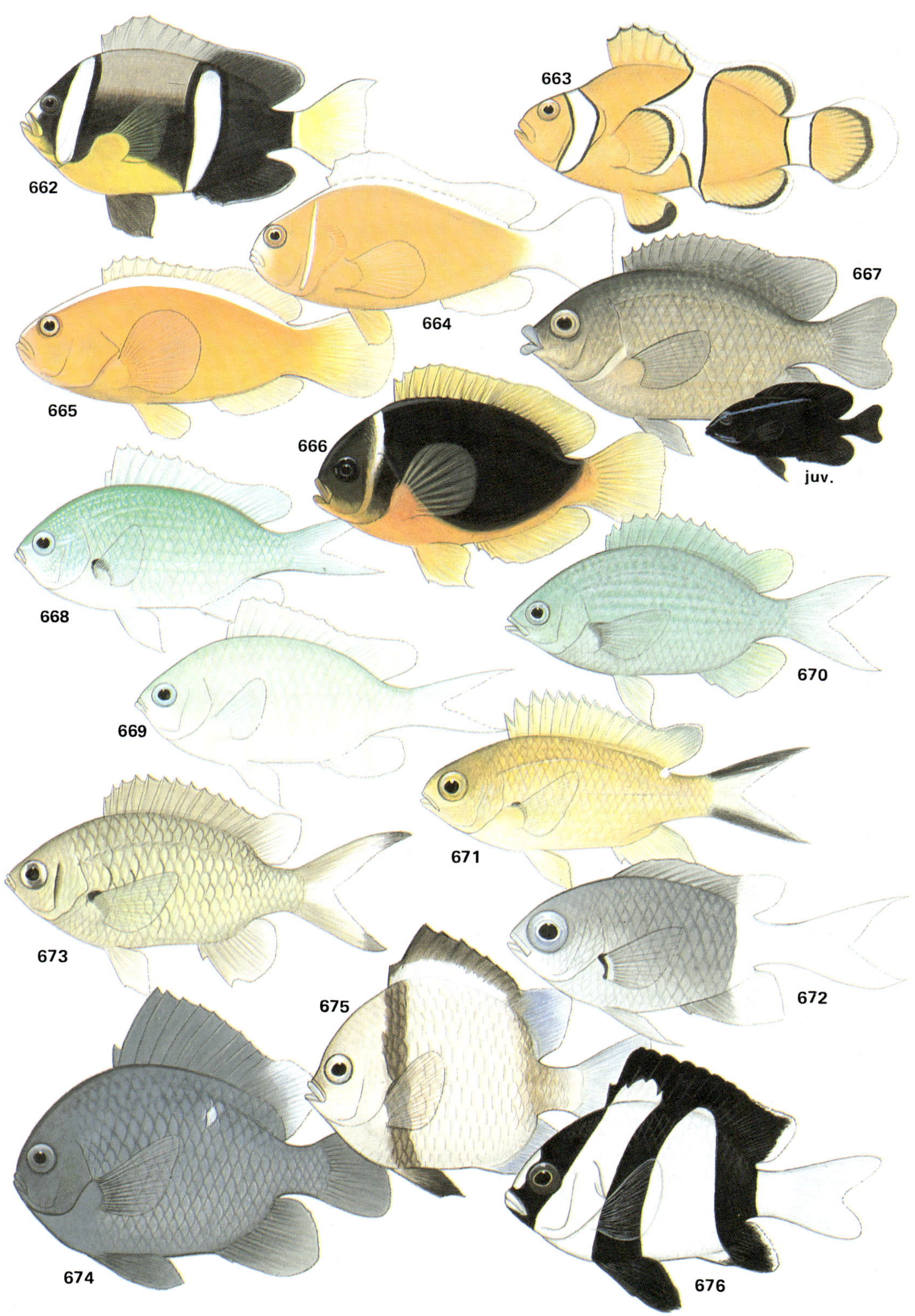

677 DICK'S DAMSEL
Plectroglyphidodon dickii (Liénard)
Inhabits coral reefs, frequently associated with *Acropora* coral heads; distinguished by intense blackish bar on rear part of body; Abrolhos northwards; Indo-C. Pacific; to 10 cm.

678 JOHNSTON DAMSEL
Plectroglyphidodon johnstonianus Fowler & Ball
Inhabits coral reefs, frequently associated with *Acropora* or *Pocillopora* coral heads; similar to 678, but posterior blackish area broader and more diffuse; Abrolhos northwards; Indo-C. Pacific; to 9 cm.

679 JEWEL DAMSEL
Plectroglyphidodon lacrymatus (Quoy & Gaimard)
Inhabits coral reefs, a territorial species that consumes algae; distinguished by numerous bright blue spots on the head and body; Abrolhos northwards; Indo-C. Pacific; to 10 cm.

680 WHITE-BANDED DAMSEL
Plectroglyphidodon leucozona (Bleeker)
Inhabits coral reefs exposed to wave action; distinguished by broad white bar on middle of sides; Abrolhos northwards; Indo-C. Pacific; to 12 cm.

681 ALEXANDER'S DAMSEL
Pomacentrus alexanderae Evermann & Seale
Inhabits coral reefs; distinguished by light body colour and black pectoral fin base; Dampier Archipelago northwards; W. Pacific and E. Indian Ocean; to 11 cm.

682 AMBON DAMSEL
Pomacentrus amboinensis Bleeker
Inhabits coral reefs, often in sandy areas; distinguished from 685 by its preference for open, sandy habitats, less brilliant yellow colour, and slightly larger spot on the upper pectoral fin base; juvenile has spot at rear of dorsal fin; Dampier Archipelago northwards; W. Pacific and E. Indian Ocean; to 11 cm.

683 NEON DAMSEL
Pomacentrus coelestis Jordan & Starks
Inhabits coral reefs, frequently amongst dead coral rubble; distinguished by bright blue colouration; Rottnest Island northwards; Indo-C. Pacific; to 9 cm.

684 MILLER'S DAMSEL
Pomacentrus milleri Taylor
Inhabits coral and rocky reefs, usually near shore; distinguished by plain brownish or grey colour and 14 dorsal spines (other similar species have 12 or 13); juveniles mainly yellow with blue stripes on back; Rottnest Island northwards; N. Australia only; to 10 cm.

685 LEMON DAMSEL
Pomacentrus moluccensis Bleeker
Inhabits coral reefs, usually in the vicinity of live branching corals; distinguished from 682 by brighter colouration and smaller spot on the upper pectoral fin base; Point Quobba northwards; mainly W. Pacific; to 7.5 cm.

686 PRINCESS DAMSEL
Pomacentrus vaiuli Jordan & Seale
Inhabits coral reefs; distinguished by rows of spots along side of body and spot on rear part of dorsal fin; Point Quobba northwards; mainly W. and C. Pacific; to 9 cm.

687 SANDY DAMSEL
Pomacentrus nagasakiensis Tanaka
Inhabits coral reefs, usually found in sandy areas around coral or rock outcrops; distinguished by black pectoral base, pale tail, and faint, wavy stripes on tail and rear part of dorsal and anal fins; juveniles are mainly bluish; Point Quobba northwards; W. Pacific and E. Indian Ocean; to 10 cm.

688 PACIFIC GREGORY
Stegastes fasciolatus (Ogilby)
Inhabits coral reefs, usually in areas exposed to moderate wave action; distinguished by overall brownish colour and lavender spots on cheek and belly; Point Quobba northwards; Indo-C. Pacific; to 13 cm.

689 BLUNT-SNOUT GREGORY
Stegastes lividus (Bloch & Schneider)
Inhabits coral reefs, usually amongst dead staghorn coral; a very pugnacious fish that continually chases other fishes (and divers) from its territory; distinguished from the similar 688 by the blunter snout and much wider gap between the eye and mouth; colour is variable, some fish are entirely brown; Point Quobba northwards; Indo-C. Pacific; to 15 cm.

690 WESTERN GREGORY
Stegastes obreptus Whitley
Inhabits coral and rocky reefs, sometimes in weedy areas; distinguished by overall dark brown colour; the juvenile is bright yellow with a spot on the dorsal fin; Rottnest Island northwards; W. Pacific and E. Indian Ocean; to 15 cm.

691 DUSKY GREGORY
Stegastes nigricans (Lacepède)
Inhabits coral reefs usually amongst algal covered branches of dead coral; distinguished by blackish spots at base of upper pectoral fin and last dorsal fin ray; nest guarding males assume a very different pattern (691b); Point Quobba northwards; Indo-C. Pacific; to 14 cm.

692 GULF DAMSEL
Pristotis jerdoni (Day)
Inhabits flat, sandy bottoms; often caught by trawlers; distinguished by plain colouration, slender shape, forked tail and small black spot on upper pectoral fin base; Shark Bay northwards; Indo-W. Pacifc; to 14 cm.

677
678
679
680
681
682
683
684
juv.
685
686
687
688
689
690
691b
691
juv.
692

693 THREE-BARRED BOARFISH
Histiopterus typus (Temminck & Schlegel)
Inhabits deep reefs on the continental shelf; distinguished by protruding snout, elevated dorsal fin and dark bars; Broome northwards; mainly W. Pacific; to 35 cm.

694 TWINSPOT HAWKFISH
Amblycirrhitus bimacula (Jenkins)
Inhabits coral reefs; distinguished by pale-edged dark spots on gill cover and below rear base of dorsal fin; Ningaloo Reef northwards; Indo-C. Pacific; to 8 cm.

695 BLOTCHED HAWKFISH
Cirrhitichthys aprinus (Cuvier)
Inhabits coral or rocky reefs; distinguished by irregular-shaped brown to red bars on side, brown spot at upper corner of gill cover, and filament at middle of dorsal fin; Abrolhos northwards; Indo-Australian Archipelago; to 10 cm.

696 SHARP-HEADED HAWKFISH
Cirrhitichthys oxycephalus (Bleeker)
Inhabits coral or rocky reefs; distinguished by red to brown spots or blotches and prolonged filament at middle of dorsal fin; Shark Bay northwards; Indo-E. Pacific; to 9 cm.

697 LYRETAIL HAWKFISH
Cyprinocirrhites polyactis (Bleeker)
Inhabits deeper offshore reefs, the only known W.A. specimens were from 132 m depth; distinguished by lack of marks on body, strongly forked tail, and filament at middle of dorsal fin; Barrow Island northwards; Indo-W. Pacific; to 15 cm.

698 FRECKLED HAWKFISH
Paracirrhites forsteri (Bloch & Schneider)
Inhabits coral reefs; distinguished by red to brown spots on head and dark upper half of body; Ningaloo Reef northwards; Indo-C. Pacific; to 20 cm.

699 RING-EYED HAWKFISH
Paracirrhites arcatus (Cuvier)
Inhabits coral reefs; distinguished by orange loop behind eye, 3 orange streaks on lower edge of gill cover, and white streak on sides; Shark Bay northwards; Indo-C. Pacific; to 13 cm.

700 ORNATE HAWKFISH
Paracirrhites hemistictus (Günther)
Inhabits offshore coral reefs; distinguished by dense spotting on upper half of body; offshore reefs of North West Shelf; Indo-C. Pacific; to 29 cm.

701 OBTUSE SANDFISH
Squamicreedia obtusata Rendahl
Inhabits sand bottoms; distinguished by slender body, dorsally positioned eyes, and pattern of small spots; Dirk Hartog Island northwards; N. Australia only; to 8 cm.

702 TOMMYFISH
Limnichthys fasciatus Waite
Inhabits sand bottoms, frequently buried below surface with only eyes protruding; distinguished by tiny size, pointed snout, dark stripe along middle of side and saddle-like spots on back; entire coast of W.A.; E. Indian Ocean and W. Pacific; to 4.5 cm.

703 BANDFISH
Acanthocepola abbreviata (Valenciennes)
Inhabits sand or silt bottoms; distinguished by elongate, laterally compressed body with pointed tail; Shark Bay northwards; E. Indian Ocean and W. Pacific; to 50 cm.

704 OCELLATED EEL-BLENNY
Blennodesmus scapularis Günther
Inhabits shallow coastal reefs; similar to 705, but pale-edged dark spot above gill cover (not on it) and lacks distinct pale spots arranged in longitudinal rows; Shark Bay northwards; N. Australia only; to 10 cm.

705 SPINY EEL-BLENNY
Congrogadus spinifer Borodin
Inhabits sand bottoms with weed and sponge; similar to 704, but pale-edged dark spot on gill cover (not above it), pale stripe behind eye, and distinct pale spots on side arranged in longitudinal rows; Exmouth Gulf northwards; N. Australia only; to 13 cm.

706 CARPET EEL-BLENNY
Congrogadus subducens (Richardson)
Inhabits rock crevices in weedy areas near coral reefs; similar to 705, but much larger, bigger lips, less distinct spot on gill cover, and pale spots on side more diffuse; Shark Bay northwards; Andaman Sea and W. Pacific; to 45 cm.

707 SPOTTED EEL-BLENNY
Notograptus guttatus Günther
Inhabits turbid inshore reefs; in rocky crevices; distinguished by elongate white body with longitudinal rows of small dark spots; Ningaloo Reef northwards; N. Australia and New Guinea; to 14 cm.

708 SHARK BAY EEL-BLENNY
Notograptus gregoryi Whitley
Inhabits sand-weed areas; distinguished by general dark colouration and several dark spots on head; a rare species known from only a few examples; known thus far only from Shark Bay; to 10 cm.

HAWKFISHES, ETC.

Hawkfishes of the family Cirrhitidae (694-696) are benthic inhabitants of coral reefs. Although found in all tropical seas, most of the 35 species are confined to the Indo-Pacific region. Hawkfishes have free projecting rays on the lower part of the pectoral fins that are adapted for perching on coral branches, gorgonians, etc. They are predators of small fishes and crustaceans. Most hawkfishes are less than 15 cm in length.

Most of the remaining fishes on this plate are seldom seen due to their cryptic habits. The members of the family Creediidae (701-702) are small fishes that bury themselves in the sand. The Bandfish (703) of the family Cepolidae occurs over soft bottoms and shelters in burrows. Eel-Blennies (704-706) are usually placed in a separate family (Congrogadidae), but recent research indicates they should be grouped with the Pseudochromidae (Plate 21; 333-340). They are usually found amongst weeds and rocks. The species in the genus *Notograptus* (707-708) are similar in shape, but are classified in a separate family, the Notograptidae.

693
694
695
696
697
699
juv.
698
700
701
702
704
703
705
706
707
708

709 FLAT-TAIL MULLET
Liza argentea Quoy & Gaimard
Inhabits coastal waters including estuaries; a plain silvery species without distinguishing marks; has 35-38 scales in lateral line, lacks an enlarged pointed scale at upper pectoral fin base and has 10 soft (excluding 3 spines) anal fin rays (versus 9 or 10 for other mullets on this page); also known as Jumping mullet and Tiger mullet; mainly confined to southern waters but ranges north to Shark Bay; Australia only; to 32 cm. ★★★

710 GREENBACK MULLET
Liza subviridis (Valenciennes)
Inhabits coastal waters including estuaries; distinguished by greenish back, gelatinous membrane partially covering eye, 27-32 scales in lateral line, and lacks an enlarged pointed scale at upper pectoral fin base, also tail narrowly dark-edged; Exmouth Gulf northwards; Indo-C. Pacific; to 30 cm. ★★★

711 DIAMOND-SCALE MULLET
Liza vaigiensis (Quoy & Gaimard)
Inhabits coastal waters including estuaries; distinguished by square-shaped tail, scales on upper side with dark blotch giving appearance of stripes, and 24-27 scales in lateral line, juvenile has black pectoral fins; Exmouth Gulf northwards; Indo-C. Pacific; to 55 cm. ★★★

712 SAND MULLET
Myxus elongatus Günther
Inhabits coastal waters, frequently off beaches or in estuaries; distinguished by pointed head, black spot at upper pectoral fin base, and 43-46 scales in lateral line; also known as Tallegalane and Lano; mainly confined to southern waters, but ranges north to Shark Bay; mainly W. Pacific; to 42 cm. ★★★

713 SEA MULLET
Mugil cephalus Linneaus
Inhabits coastal waters, entering estuaries and fresh water; distinguished by flattened head, gelatinous membrane covering eye, an enlarged pointed scale at upper pectoral fin base, 38-42 scales in lateral line, and diffuse stripes on side; entire coast of W.A.; worldwide temperate and tropical seas; to 79 cm. ★★★

714 BLUE-TAIL MULLET
Valamugil buchanani (Bleeker)
Inhabits coastal waters including estuaries; distinguished by forked tail, gelatinous membrane around rim of eye, an enlarged pointed scale at upper pectoral fin base, and 32-35 scales in lateral line; Exmouth Gulf northwards; Indo-W. Pacific; to 40 cm. ★★★

715 NORTHERN THREADFIN
Polydactylus plebius Broussonet
Inhabits coastal waters, frequently off beaches; distinguished by divided pectoral fin with lower part containing 5 free filamentous rays, and narrow stripes on side; also known as Striped salmon; Shark Bay northwards; Indo-C. Pacific; to 45 cm. ★★★

716 GUNTHER'S THREADFIN
Polydactylus multiradiatus Günther
Inhabits coastal waters over sand or mud bottoms; distinguished by divided pectoral fin with lower part containing 7 free filamentous rays and most of first dorsal fin blackish; Broome northwards; mainly W. Pacific; to 20 cm. ★★★

717 BLACK-FINNED THREADFIN
Polydactylus nigripinnis Munro
Inhabits coastal waters over sand or mud bottoms; distinguished by divided pectoral fin with lower part containing 6 free filaments and upper part black; Derby northwards; N. Australia and New Guinea; to 20 cm.

718 GIANT THREADFIN
Eleutheronema tetradactylum (Shaw)
Inhabits coastal waters over sand or mud bottoms, sometimes in estuaries; distinguished by divided pectoral fin with lower part containing 4 free filaments and lips absent except for small section at rear of lower jaw; also known as Cooktown salmon and Rockhampton kingfish; Port Hedland northwards; Indo-W. Pacific; to 120 cm. ★★★

719 MILITARY SEAPIKE
Sphyraena putnamiae Jordan & Seale
Inhabits coastal waters; distinguished by chevron-shaped markings on side and forked tail; Onslow northwards; Indo-W. Pacific; to at least 50 cm. ★★

720 BARRACUDA
Sphyraena barracuda (Walbaum)
Inhabits coastal waters and offshore reefs; distinguished by large size, dark or dusky fins, diffuse dark bars on back, frequently with scattered dark blotches on side (not shown) and truncate tail, 721 is similar, but more slender and has smaller scales (about 123-135 in lateral line versus 75-90); a curious fish that sometimes follows divers; has been known to attack humans in the Atlantic, but no Australian incidents have been reported; Albany northwards; worldwide in tropical seas; to 180 cm. ★

721 GIANT SEAPIKE
Sphyraena jello Cuvier
Inhabits coastal waters and offshore reefs; similar to 720, but more slender and has smaller scales (see above); Ningaloo Reef northwards; Indo-W. Pacific; to 125 cm. ★

722 STRIPED SEAPIKE
Sphyraena obtusata Cuvier
Inhabits sand-weed areas, often in the vicinity of rocky or coral reefs; distinguished by pair of dusky yellowish stripes on side and yellow tail; Albany northwards; Indo-C. Pacific; to 55 cm. ★★

MULLETS, THREADFINS AND BARRACUDAS

Mullets (709-714), Threadfins (715-718), and Barracudas (719-722) are placed together in the suborder Mugiloidei. They are primarily inhabitants of tropical and subtropical seas and estuaries, with a few species also penetrating the lower parts of freshwater streams. Mullets (Mugilidae; worldwide about 75 species) are mainly algal feeders that have very small teeth or may lack them entirely. The largest species is about 90 cm TL, but most are under about 40 cm. Barracudas (Sphyraenidae; worldwide about 18 species) are found near coastal reefs. They have large fang-like teeth and are known to occasionally attack humans. The largest species reaches about 2 m TL. Threadfins (Polynemidae; worldwide about 35 species) inhabit sand or mud bottoms.

723 SPOTTED CHISEL-TOOTHED WRASSE
Anampses caeruleopunctatus Rüppell
Inhabits coral reefs and rocky areas; females distinguished by blue spotting on head and body, male by greenish colour with blue streak on each scale and yellowish bar at level of pectoral fins; Rottnest Island northwards; Indo-C. Pacific; to 30 cm. ★★

724 SCRIBBLED CHISEL-TOOTHED WRASSE
Anampses geographicus Valenciennes
Inhabits coastal reefs, frequently in weedy areas; female distinguished by large spot at rear of dorsal and anal fins, male by blue lines on head and numerous blue spots or streaks on sides; Albany northwards; mainly W. Pacific; to 31 cm. ★★

725 BLUE AND YELLOW WRASSE
Anampses lennardi Scott
Inhabits coral reefs; female distinguished by blue and yellow stripes, male by yellowish patch above pectoral fin and blue streaks on tail; Ningaloo Reef northwards; N.W. Australia only; to 28 cm. ★★

726 YELLOWTAIL WRASSE
Anampses meleagrides Valenciennes
Inhabits coral reefs; female distinguished by white spots and yellow tail, male by dark area forming "pointed" tail lobes, Ningaloo Reef northwards; Indo-C. Pacific; to 22 cm. ★★

727 CORAL PIGFISH
Bodianus axillaris (Bennett)
Inhabits coral reefs; distinguished by large black spot on dorsal and anal fins, also at base of pectoral fin; Abrolhos northwards; Indo-C. Pacific; to 20 cm. ★★

728 SADDLEBACK PIGFISH
Bodianus bilunulatus (Lacepède)
Inhabits coral reefs; distinguished by black saddle below rear of dorsal fin, juveniles with large black area at rear of body; Abrolhos northwards; Indo-W. Pacific; to 55 cm. ★★★

729 GOLDSPOT PIGFISH
Bodianus perditio (Quoy and Gaimard)
Inhabits the vicinity of coral and rocky reefs, often over sand or rubble in deeper water; distinguished by yellow patch on middle of back followed by blackish area; Abrolhos northwards; Indo-W. Pacfic; to 53 cm; 3.12 kg. ★★★

730 BLACKSPOT PIGFISH
Bodianus vulpinus (Richardson)
Inhabits mainly rocky reefs; female distinguished by narrow stripes and sometimes dark blotches on sides, male (see plate 70) by black blotch at base of middle dorsal spines and reddish margin on upper and lower edges of tail; Cape Naturaliste to Shark Bay; mainly W. Pacific; to 60 cm; 1.8 kg. ★★★★

731 SCARLET-BREASTED MAORI WRASSE
Cheilinus fasciatus (Bloch)
Inhabits coral reefs; distinguished by red area at front of body and prominent dark bars; North West Shelf; Indo-W. Pacific; to 38 cm. ★★★

732 YELLOW-DOTTED MAORI WRASSE
Cheilinus chlorurus (Bloch)
Inhabits inshore reefs, in both coral and weed areas; distinguished by mottled pattern with small pale spots on sides; Point Quobba northwards; Indo-C. Pacific; to 45 cm; .9 kg. ★★★

733 WHITEBAND MAORI WRASSE
Cheilinus unifasciatus Streets
Inhabits coral reefs; distinguished by white band across tail base; Ningaloo Reef northwards; mainly W. and C. Pacific; to 30 cm. ★★

734 TRIPLETAIL MAORI WRASSE
Cheilinus trilobatus Lacepède
Inhabits coral reefs; distinguished by numerous pale spots on head, elongated rear part of dorsal and anal fins, and irregular outline of tail; Coral Bay northwards; Indo-C. Pacific; to 45 cm; 1.0 kg. ★★★

WRASSES

The colourful wrasses of the family Labridae (Plates 47-51) are one of the most conspicuous fish groups inhabiting tropical coral reefs. The family is also well represented in warmer temperate seas, for example along the southern coastline of Australia. There are an estimated 400 species including some which are still undescribed. The family is extremely diverse regarding colours, shape, behaviour, and ecological preferences. Most species live over sand, rubble, weed, or coral and rock substrata. Wrasses occur over a wide depth range including shallow tidal pools to depths of at least 100 m. They are diurnally active, feeding on a wide variety of benthic and pelagic invertebrates. At dusk some species bury themselves in the sand where they 'sleep' through the night. Razorwrasses (*Xyrichtys*) also dive under the sand when threatened. Female to male sex reversal is common in many species. Dramatic colour changes sometimes occur during growth, with juveniles, females and males exhibiting different patterns. Most species are medium-sized (about 20-40 cm), although the Humphead Wrasse (*Cheilinus undulatus*) which has never been recorded from coastal W.A. (but present at Rowley Shoals), grows to about 2 m. The larger species are often good eating.

735 SHARP-NOSED WRASSE
Cheilio inermis (Forsskål)
Inhabits weed beds; distinguished by elongate shape, males have blackish patch behind pectoral fins, females frequently display a narrow dark stripe along middle of sides; Point Quobba northwards; Indo-C. Pacific; to 50 cm; .33 kg.

736 PEACOCK WRASSE
Cirrhilabrus temmincki Bleeker
Inhabits outer coral reefs, usually where there is loose rubble; male distinguished by red colour on upper half of head and body and very elongate pelvic fin filaments, females less vivid and lack pelvic filaments; Abrolhos to North West Cape; mainly W. Pacific; to 10 cm.

737 BLUE TUSKFISH
Choerodon cyanodus (Richardson)
Inhabits coral reefs and flat bottoms; distinguished by white chin, white spot on middle of back and scribble markings on tail; Exmouth Gulf northwards; N. Australia only; to 60 cm; 6.65 kg. ★★★

738 PURPLE TUSKFISH
Choerodon cephalotes (Castelnau)
Inhabits reefs and flat bottoms; distinguished by dark colour on upper half and light colour on lower half with whitish patch above pectoral fin; Port Hedland northwards; N. Australia only; to 30 cm; .365 kg. ★★★

739 BALDCHIN GROPER
Choerodon rubescens (Günther)
Inhabits coral reefs and rock-weed areas; distinguished by abruptly pale chin and pale area at base of pectoral fins, the head profile becomes increasingly steep with growth; Geographe Bay to Coral Bay; W.A. only; to 90 cm; 7.1 kg. ★★★★

740 JORDAN'S WRASSE
Choerodon jordani (Snyder)
Inhabits sandy areas adjacent to reefs; distinguished by blackish wedge on rear part of body and large white spot below end of dorsal fin; Abrolhos northwards; mainly W. Pacific; to 13 cm.

741 REDSTRIPE TUSKFISH
Choerodon vitta Ogilby
Inhabits flat sandy or weedy areas; distinguished by reddish stripe along middle of sides; Exmouth Gulf northwards; N. Australia only; to 20 cm. ★★

742 BLACKSPOT TUSKFISH
Choerodon schoenleinii (Valenciennes)
Inhabits sand and weed areas adjacent to coral reefs; distinguished by overall bluish colour and black spot at base of middle of dorsal fin; Point Quobba northwards; mainly W. Pacific; to 80 cm; 9.0 kg. ★★★

743 BLUESPOTTED TUSKFISH
Choerodon cauteroma Gomon & Allen
Inhabits sand and weed areas adjacent to coral reefs; distinguished by yellow colouration and dark streak below middle of spiny dorsal fin; Shark Bay northwards; N.W. Australia only; to 24 cm. ★★

744 WEDGE-TAILED TUSKFISH
Choerodon sugillatum Gomon
Inhabits flat sandy or weedy areas; distinguished by blue band at pectoral fin base and blue streak above pectoral fin; North West Shelf; N.W. Australia only; to 24 cm. ★★

745 DARKSPOT TUSKFISH
Choerodon monostigma (Ogilby)
Inhabits flat sandy or weedy areas; distinguished by prominent black spot at middle of dorsal fin and striped tail; Port Hedland northwards; N. Australia only; to 25 cm. ★★

746 ZAMBOANGA TUSKFISH
Choerodon zamboangae (Seale & Bean)
Inhabits flat sandy or weedy areas; distinguished by dark back and wedge-shaped reddish mark; North West Shelf; mainly W. Pacific; to 25 cm. ★★

747 RED-FINNED RAINBOWFISH
Coris gaimardi (Quoy & Gaimard)
Inhabits sandy areas adjacent to coral reefs; distinguished by bright colour pattern and elongate spines at front of dorsal fin, juveniles bright red with white saddles; Point Quobba northwards; Indo-C. Pacific; to 30 cm.

748 HUMP-HEADED WRASSE
Coris aygula Lacepede
Inhabits rubble, weed and sandy areas adjacent to reefs; male distinguished by hump on forehead and ragged tail margin, females by bicolour pattern with spots on the head and fins, juveniles with twin "eye-spots"; Rottnest Island northwards; Indo-C. Pacific; to 100 cm. ★★★

749 BLACK-STRIPED WRASSE
Coris pictoides Randall & Kuiter
Inhabits rubble, sand, or weedy areas adjacent to coral reefs; distinguished by black stripes on upper sides; Dampier Archipelago northwards; W. Pacific and E. Indian Oceans; to 11 cm.

750 SPOTTED-TAIL WRASSE
Coris caudimacula (Quoy & Gaimard)
Inhabits sand or weedy areas adjacent to coral reefs; distinguished by dark band or blotches along upper side, broken bands or spots on tail, and usually a prominent "ear-spot" at rear edge of gill cover; Shark Bay northwards; Indian Ocean; to 20 cm.

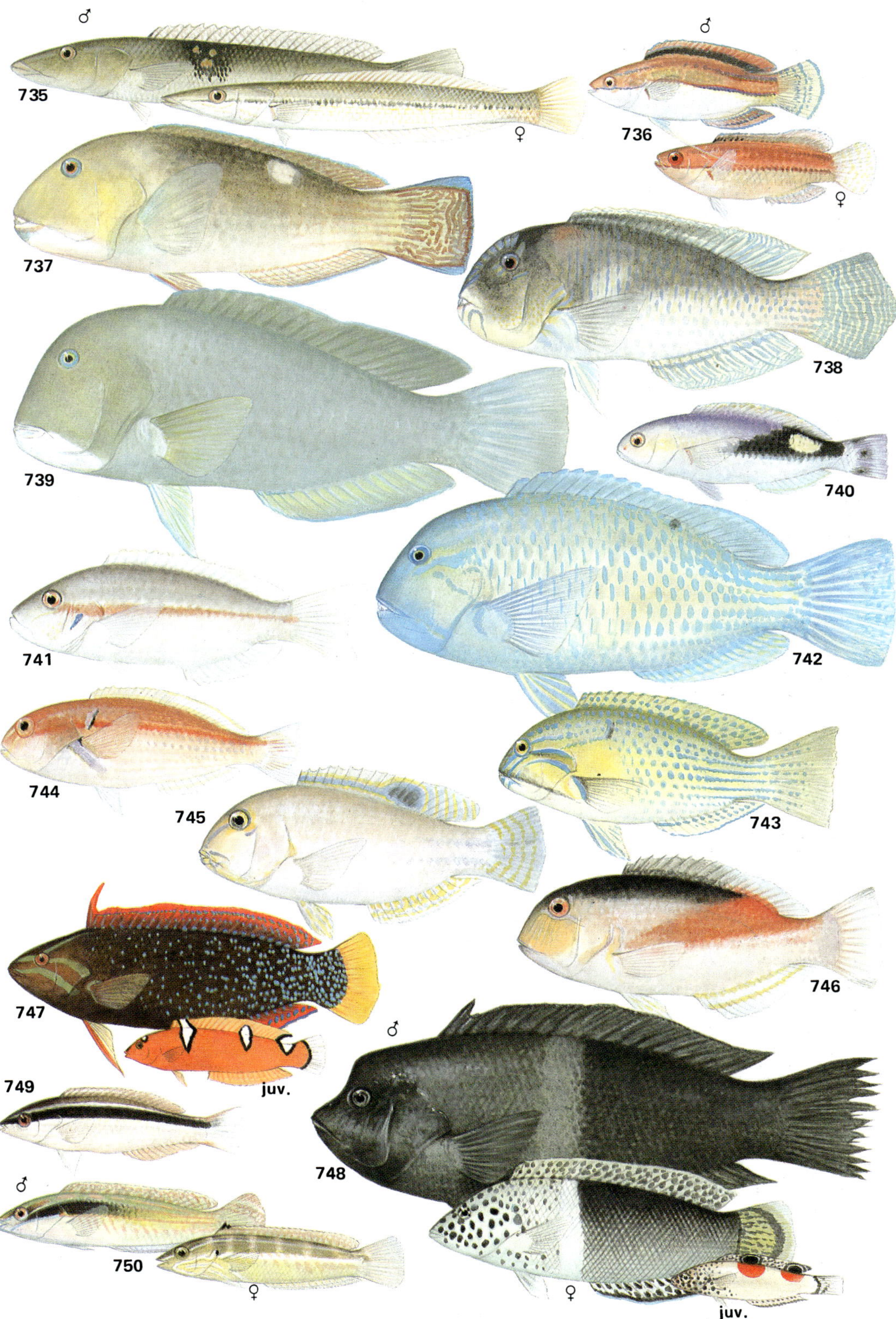

751 SLINGJAW WRASSE
Epibulus insidiator (Pallas)
Inhabits coral reefs; distinguished by its highly protrusible jaw; two colour varieties are commonly encountered, one that is largely bright yellow and a dark variety; Abrolhos northwards; Indo-C. Pacific; to 35 cm. ★★

752 CLUBNOSED WRASSE
Gomphosus varius (Lacepède)
Inhabits coral reefs; distinguished by its elongate snout, females are generally lighter in colour than males; Abrolhos northwards; Indo-C. Pacific; to 28 cm.

753 FOURSPOT WRASSE
Halichoeres hortulanus (Lacepède)
Inhabits coral reefs; female distinguished by blackish area below spiny dorsal fin, and ocellus on middle of dorsal fin, male by yellow spot or broad pale bar behind head; Point Quobba northwards; Indo-C. Pacific; to 26 cm.

754 RED-LINED WRASSE
Halichoeres biocellatus Schultz
Inhabits coral reefs; distinguished by series of red streaks and lines on head and sides; Point Quobba northwards; mainly W. Pacific; to 15 cm.

755 SPECKLED RAINBOWFISH
Halichoeres marginatus Rüppell
Inhabits coral reefs, distinguished by dark green-brown colour with darker stripes (most apparent anteriorly) and yellow margin on tail, young have striped pattern with an ocellus on the dorsal fin; Shark Bay northwards; Indo-C. Pacific; to 18 cm.

756 THREESPOT WRASSE
Halichoeres trimaculatus (Griffith)
Inhabits sand and rubble flats adjacent to coral reefs; female distinguished by overall pale colour and spot on upper tail base, male is more ornate, but also has prominent spot on upper tail base; Abrolhos northwards; Indo-C. Pacific; to 25 cm.

757 PURPLE WRASSE
Halichoeres melanochir Fowler & Bean
Inhabits coral reefs; distinguished by overall purple-brown colour and yellow-orange pelvic fins, young specimens have 2 spots on the dorsal fin and a third spot on the upper tail base; Point Quobba northwards; mainly W. Pacific; to 10 cm.

758 DIAMOND WRASSE
Halichoeres nigrescens Bleeker
Inhabits rubble and weed areas near coral reefs; distinguished by 4-5 diffuse, broad dark bars on upper two-thirds of side with yellow areas in between and dark spot at pectoral fin base; Dampier Archipelago northwards; Indo-Australian Archipelago; to 13 cm.

759 SADDLED RAINBOWFISH
Halichoeres margaritaceous (Valenciennes)
Inhabits rubble and weed areas near coral reefs; similar to 760, but has the bands running across the cheek rising posteriorly (/) and usually has 13 pectoral-fin rays; Point Quobba northwards; Indo-C. Pacific; to 12 cm.

760 NEBULOUS WRASSE
Halichoeres nebulosus (Valenciennes)
Inhabits rubble and weed near coral reefs; similar to 759 but has the bands running across the cheek descending posteriorly (\) and usually has 14 pectoral-fin rays; Point Quobba northwards; Indo-W. Pacific; to 12 cm.

761 FIVE-BANDED WRASSE
Hemigymnus fasciatus (Bloch)
Inhabits coral reefs; distinguished by fleshy lips and broad dark bands on sides; Rottnest Island northwards; Indo-C. Pacific; to 80 cm. ★★★

762 THICK-LIPPED WRASSE
Hemigymnus melapterus (Bloch)
Inhabits coral reefs; distinguished by large size and fleshy lips, smaller individuals (up to about 40-50 cm have characteristic 'bicolour' pattern (pale anteriorly and dark posteriorly); Shark Bay northwards; Indo-C. Pacific; to 90 cm. ★★★

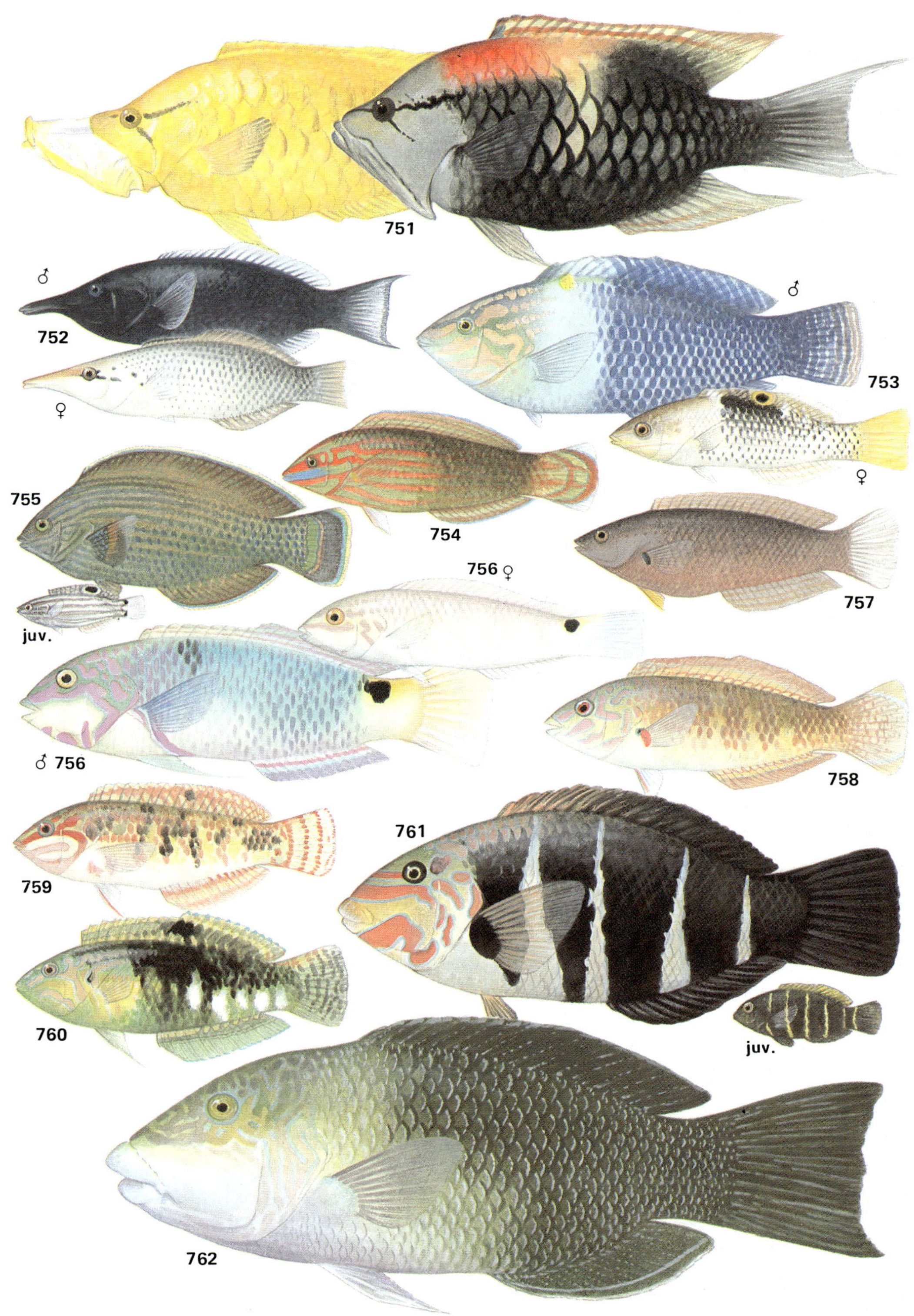
751
752
753
754
755
juv.
756 ♀
♂ 756
757
758
759
760
761
juv.
762
♂
♀
♂
♀

763 RINGED SLENDER WRASSE
Hologymnosus annulatus (Lacepède)
Inhabits coral reefs, often seen over sand or rubble areas; male distinguished by series of narrow dark bars on side, these also visible in female, which has a more slender body, juveniles black except golden yellow on back and top of head; Shark Bay northwards; Indo-W. Pacific; to 40 cm; .347 kg.　★★

764 BICOLOR CLEANERFISH
Labroides bicolor Fowler & Bean
Inhabits coral reefs; feeds on ectoparasites from other fishes; distinguished from 765 by light coloured tail region; Ningaloo Reef northwards; Indo-C. Pacific; to 14 cm.

765 CLEANERFISH
Labroides dimidiatus (Valenciennes)
Inhabits coral reefs; feeds on ectoparasites from other fishes; distinguished from 764 by blackish stripe running through eye and continuing to tail; Rottnest Island northwards; Indo-C. Pacific; to 12 cm.

766 ONE-LINED WRASSE
Labrichthys unilineatus (Guichenot)
Inhabits coral reefs, frequently seen amongst branching *Acropora* coral; distinguished by thin blue lines on side and male has pale bar behind head, juveniles have a single pale stripe along the middle of the side; Ningaloo Reef northwards; Indo-W. Pacific; to 17 cm.

767 WEEDY WRASSE
Leptojulis cyanopleura (Bleeker)
Inhabits weed beds in the vicinity of coral reefs, male distinguished by reddish fins, black spot above pectoral fin and oblique red stripes on tail, female by dark stripe along middle of side; Point Quobba northwards; Indo-W. Pacific; to 12 cm.

768 SIX-LINED WRASSE
Pseudocheilinus hexataenia (Bleeker)
Inhabits coral reef crevices; distinguished by 6 red-orange stripes on side and small black spot on upper edge of tail base; North West Shelf; Indo-C. Pacific; to 8 cm.

769 FLAGFIN WRASSE
Pterogogus enneacanthus (Bleeker)
Inhabits weedy patches on coral reefs; distinguished by free filamentous spines at front of dorsal fin and prominent spot on gill cover; Point Quobba northwards; Indo-W. Pacific; to 20 cm.

770 ORNATE WRASSE
Macropharyngodon ornatus Randall
Inhabits coral reefs; female distinguished by pale spots in rows on side and reddish fins with light markings, males similar, but generally darker; Ningaloo Reef northwards; Indo-W. Pacific; to 11 cm.

771 BLACK LEOPARD WRASSE
Macropharyngodon negrosensis Herre
Inhabits coral reefs; female distinguished by blackish colour with blue-green scale margins and upper and lower edges of tail with dark streaks; Ningaloo Reef northwards; mainly W. Pacific; to 12 cm.

772 REDSPOT WRASSE
Stethojulis bandanensis (Bleeker)
Inhabits coral and rocky reefs, often in weed or rubble areas; distinguished by red patch above pectoral fin base; male has curved blue stripe on cheek and female with pale spotting on back; Rottnest Island northwards; Indo-C. Pacific; to 15 cm.

773 SILVER-STREAKED WRASSE
Stethojulis strigiventer (Bennett)
Inhabits weedy areas in the vicinity of coral and rocky reefs; female distinguished by narrow stripes on ventral half of body, male similar to 772, but lacks blue stripe on the cheek; Rottnest Island northwards; Indo-W. Pacific; to 15 cm.

774 SOELA WRASSE
Suezichthys soelae Russell
Inhabits sandy areas in 50-80 m; distinguished by dark spots on tail; North West Shelf; N.W. Australia only; to 10 cm.

775 PEARLY RAINBOWFISH
Xenojulis margaritaceous (Macleay)
Inhabits weedy areas in the vicinity of coral reefs; distinguished by dorsal fin shape (short in front and elevated posteriorly) and mottled pattern with pearly or pinkish spots on side; Point Quobba northwards; mainly W. Pacific; to 15 cm.

FISH SERVICE STATIONS

The Cleaner wrasses (764-765) are small, colourful reef inhabitants which render an important service to other fishes. They remove parasites from the body, mouth cavity, and gill chambers of numerous species, large and small. Cleaning 'stations' occur at regular intervals on coral reefs and research has shown they are an integral component in promoting good 'health' in the overall fish community. Each station is occupied by one or more cleaners. The demand for their services is readily apparent as several fishes often 'queue' while awaiting their turn. The cleaners show no hesitation in entering the mouth of large voracious predators such as moray eels and gropers. The False Cleanerfish (Plate 55; 826) is a member of the blenny family that is cleverly disguised as a Cleaner Wrasse. Its colour pattern, shape, and swimming motion are nearly a perfect match of those of 765. It uses its disguise to safely approach other fishes, but instead of removing parasites it quickly dashes in when at close range and rips a chunk of skin, flesh, or fin with its enlarged fangs.

763
juv.
♂
♀
764
765
766
juv
767
♂
♀
768
769
770
♂
771
♀
772
♂
♀
773
♂
♀
774
775

776 RED AND GREEN WRASSE
Thalassoma purpureum (Forsskål)
Inhabits rocky reefs and inshore coral reefs, usually where there is wave action; male distinguished by ornate pattern highlighted by reddish pink to purple stripes, female and juvenile greenish with double row of elongate red to brown blotches or stripes on middle and lower side; Rottnest Island northwards; Indo-C. Pacific; to 40 cm; .756 kg. ★★

777 BLUE-HEADED WRASSE
Thalassoma amblycephala (Bleeker)
Inhabits coral and rocky reefs, usually several females seen with each male; male distinguished by bluish head with broad pale band just behind head, female has dark stripe or is overall brownish on upper half of body and white below; Rottnest Island northwards; Indo-W. Pacific; to 15 cm.

778 SEVEN-BANDED WRASSE
Thalassoma septemfasciata Scott
Inhabits rocky and weedy reefs; male distinguished by uniform dark body and yellow pectoral fins, female by broad greenish bars on side and yellow colour on pectoral fin base; Rottnest Island to Coral Bay; W.A. only; to 31 cm; .51 kg. ★★

779 MOON WRASSE
Thalassoma lunare (Linnaeus)
Inhabits coral and rocky reefs; distinguished from 780 by magenta central portion of pectoral fin; juveniles (not shown) have a bluish belly and large black spot at base of tail; Rottnest Island northwards; Indo-W. Pacific; to 30 cm; .13 kg. ★★

780 GREEN MOON WRASSE
Thalassoma lutescens (Lay & Bennett)
Inhabits coral and rocky reefs; male similar to 779 but with blue-edged pectoral fin and lighter body colour, juvenile (not shown) with black midlateral stripe; Rottnest Island northwards; Indo-C. Pacific; to 30 cm; .375 kg. ★★

781 SIX-BANDED WRASSE
Thalassoma hardwickei (Bennett)
Inhabits coral reefs; distinguished from 782 by narrower bars some of which extend to belly region; Point Quobba northwards; Indo-C. Pacific; to 18 cm.

782 JANSEN'S WRASSE
Thalassoma janseni (Bleeker)
Inhabits coral reefs; similar to 781 but has broader bars that do not extend onto belly; Shark Bay northwards; Indo-W. Pacific; to 18 cm.

783 LONG GREEN WRASSE
Pseudojuloides elongatus Ayling & Russell
Inhabits weed beds; distinguished by elongate shape, male has orange blotch at pectoral base and blue spots on upper side, female uniform greenish; Abrolhos to Dampier Archipelago; Australlia, New Zealand, and Japan; to 15 cm.

784 BLUE-TOOTHED TUSKFISH
Xiphocheilus typus Bleeker
Inhabits flat sandy bottoms or rubble; distinguished by blue and yellow stripes in front of eye and yellowish fins; North West Shelf; Indo-Australian Archipelago; to 14 cm.

785 PAVO RAZORFISH
Xyrichtys pavo Valenciennes
Inhabits sand bottoms; uses keeled forehead to bury into the sand when threatened; distinguished by broad brown bars and antenna-like first dorsal fin; North West Shelf; Indo-C. Pacific; to 35 cm.

786 BLACKSPOT RAZORFISH
Xyrichtys dea Temminck & Schlegel
Inhabits sand bottoms; buries in sand when threatened; distinguished by antenna-like first dorsal fin, red or pinkish colour with vague bars, and small black spot below front of dorsal fin, juveniles variable, black to pale in colour; North West Shelf; mainly W. Pacific; to 35 cm.

787 CARPET WRASSE
Novaculichthys taeniurus (Lacepède)
Inhabits rubble and weedy areas near coral reefs; distinguished by diagonal stripes on cheek, speckled fins and white bar across base of tail, juvenile has elongate spines at front of dorsal fin; Dampier Archipelago northwards; Indo-C. Pacific; to 25 cm.

788 DOUBLE-HEADED PARROTFISH
Bolbometopon muricatum (Valenciennes)
Inhabits coral reefs; distinguished by large size and hump on forehead; the largest of all parrotfishes; Ningaloo Reef northwards; Indo-C. Pacific; to 130 cm. ★★★

789 RED-SPECKLED PARROTFISH
Cetoscarus bicolor (Rüppell)
Inhabits coral reefs, often in pairs; male distinguished by orange-red to pink scale margins and spots on head and front part of body, and pale stripe from mouth to anal fin, females by dense black spotting on lower two-thirds of side; juvenile mainly white with broad orange bar on head; Ningaloo Reef northwards; Indo-C. Pacific; to 50 cm. ★★★

790 SPINYTOOTH PARROTFISH
Calotomus spinidens (Quoy & Gaimard)
Inhabits weed and seagrass beds; distinguished by marbled pattern of green and brown, differs from other W.A. parrotfishes in having separate teeth (not fused to form beak-like structure); can change colour rapidly to blend with surroundings; Ningaloo Reef northwards; Indo-C. Pacific; to 19 cm.

791 BLUE-SPOTTED PARROTFISH
Leptoscarus vaigiensis (Quoy & Gaimard)
Inhabits weed and seagrass beds; similar to 790, but has fused teeth, male has pale stripe along sides and small blue spots on head, female mottled and spotted with whitish and dark brown; Rottnest Island northwards; Indo-C. Pacific; to 38 cm.

792 LONG-NOSED PARROTFISH
Hipposcarus longiceps (Valenciennes)
Inhabits coral reefs, sometimes seen over sand or rubble bottoms, often forms schools; distinguished by pointed head, female has yellowish fins and is overall pale, male bluish on sides and on snout with blue margins on most fins; Ningaloo Reef northwards; Indo-C. Pacific; to 45 cm. ★★★

793 SIX-BANDED PARROTFISH
Scarus frenatus Lacepéde
Inhabits coral reefs, usually occurring in small groups; male distinguished by abruptly lighter areas on lower half of head and posterior part of body, female by series of 6-7 dark stripes on side and reddish fins; Ningaloo Reef northwards; Indo-C. Pacific; to 40 cm. ★★★

794 SADDLED PARROTFISH
Scarus dimidiatus Bleeker
Inhabits coral reefs, usually alone or in small groups; male distinguished by dusky band from eye across gill cover with white band below, female by 3 saddle-like bars on back; Ningaloo Reef northwards; Indo-C. Pacific; to 34 cm. ★★★

795 BLUE-BARRED PARROTFISH
Scarus ghobban Forsskål
Inhabits coral reefs, seagrass and weed areas, and deeper offshore trawling grounds; male distinguished by broad blue scale margins, 3 dark streaks behind and below eye, and short blue stripe on chin, female by overall yellowish colour with diffuse blue bars and spots on side; Rottnest Island northwards; Indo-E. Pacific; to 100 cm; 6.4 kg. ★★★

796 CHAMELEON PARROTFISH
Scarus chameleon Choat & Randall
Inhabits coral reefs; male distinguished by dark stripe from snout, passing below eye across gill cover, another short stripe behind eye and vertical band above eye, also usually has large oval-shaped pale area occupying most of lower side, female generally dusky (may also have pale blotches) on upper half of side and abruptly pale below; Abrolhos northwards; mainly W. Pacific; to 28 cm.

PARROTFISHES

Parrotfishes of the family Scaridae (Plates 52-53) are closely related to the wrasses (Plates 47-51), but rather than having individual teeth in the jaws, the dental plates are fused to form a distinctive beak-like structure. This structure is well adapted for scraping algal food from the surface of the reef. Parrotfishes also ingest large amounts of coral rock and sand along with the algae and this material is ground to a fine powder by special teeth at the back of the throat. This material is then passed out with the faeces, contributing significantly to the bottom sediment. Like the wrasses, parrotfishes undergo female to male sex change and display different colour patterns according to growth stage or sex. The estimated 75 species occuring worldwide are mainly inhabitants of coral reefs, although a few species are found in weedy areas. They occur both individually or graze over the reef in large schools. At dusk parrotfishes retire to shelter in the form of crevices and ledges. Many of the species exude a strange cocoon-like mucus envelope which they remain inside during the night. This appears to be an adaptation which masks their scent, thus preventing predation by fishes such as moray eels that rely heavily on their sense of smell in locating prey. The largest parrotfishes are slightly over 1 m in length, but most species are under 50 cm.

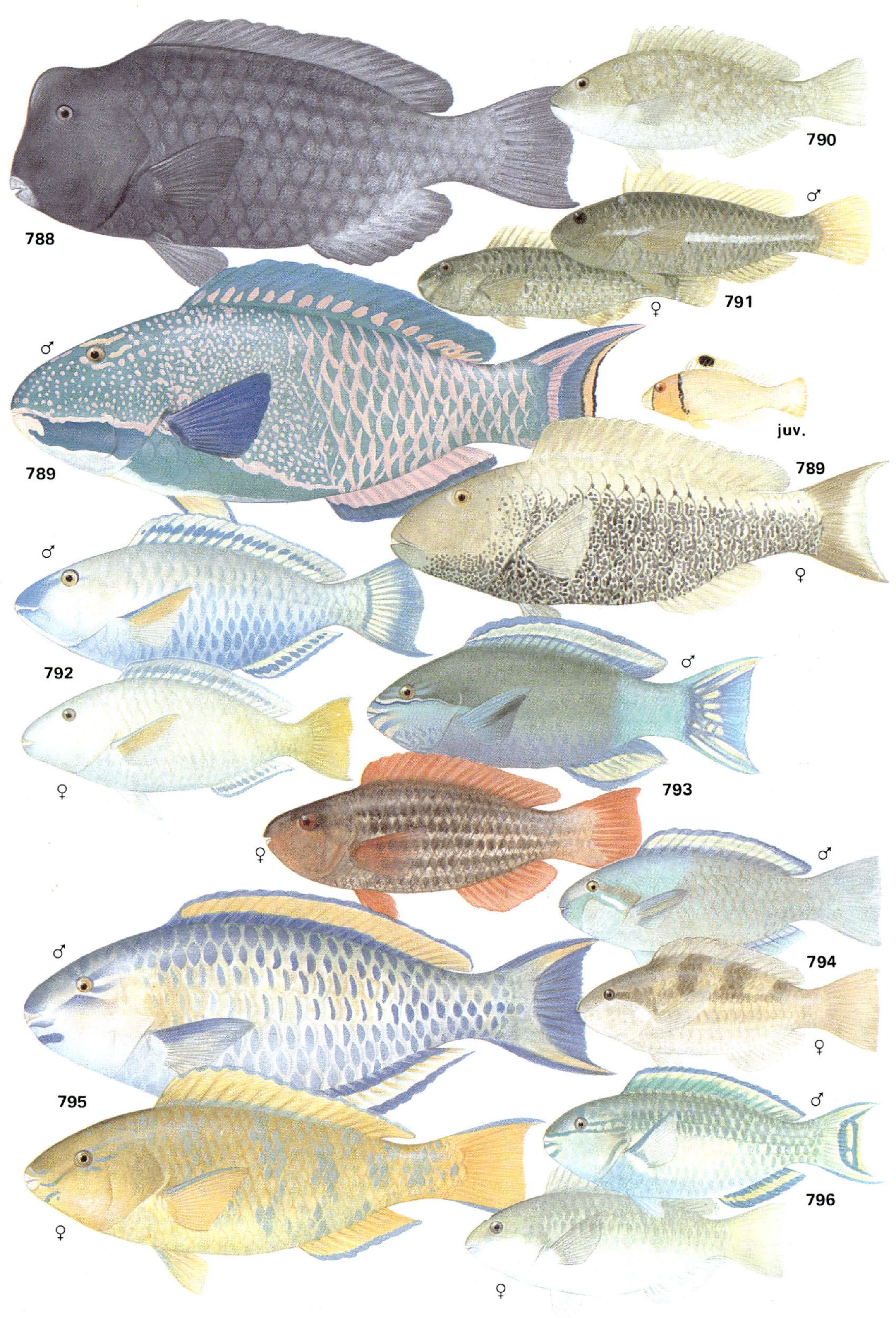
788
790
791
♂
789
♂
♀
juv.
789
♀
792
♂
793
♂
♀
794
♀
795
♂
796
♂
♀

797 STEEPHEAD PARROTFISH
Scarus gibbus Rüppell
Inhabits coral reefs and southern rocky reefs; distinguished by steep, blunt snout profile, becoming steeper with increased size, male and female similar, but 2 colour phases are exhibited, a more common green one and a red one, large terminal male sometimes mainly purple with blue area ventrally; juvenile (not shown) is dark brown with 3 white stripes on side; Rottnest Island northwards (rare south of Abrolhos); Indo-C. Pacific; to 50 cm. ★★★

798 VIOLET-LINED PARROTFISH
Scarus globiceps Valenciennes
Inhabits coral reefs; male distinguished by "maze" pattern of spots and dashes on upper back and top of head, female difficult to separate from that of 805, but often seen in company with male; Ningaloo Reef northwards; Indo-C. Pacific; to 30 cm. ★★★

799 BLUE PARROTFISH
Scarus oviceps Valenciennes
Inhabits coral reefs; male distinguished by abruptly dark area on upper part of head and adjacent portion of back, female by similar darkened area followed by a pair of light and dark patches below dorsal fin; Ningaloo Reef northwards; mainly W. and C. Pacific; to 35 cm. ★★★

800 DUSKY PARROTFISH
Scarus prasiognathus Valenciennes
Inhabits coral reefs; male distinguished by blue-green colour on head below eyes and orange colour above, female by blackish or dark brown colour, numerous small white spots on side, and reddish fins; Ningaloo Reef northwards; W. Pacific and E. Indian Ocean; to 70 cm. ★★★

801 PALENOSE PARROTFISH
Scarus psittacus Forsskål
Inhabits coral reefs; male distinguished by 2-3 blue bands behind eye and lavender or bluish-grey colour on snout and forehead, female by overall dark colour, red pelvic fins, and pale snout; Ningaloo Reef northwards; mainly W. and C. Pacific; to 30 cm. ★★★

802 EMBER PARROTFISH
Scarus rubroviolaceus Bleeker
Inhabits coral reefs; male distinguished by blunt snout and bicolour pattern, female by red colour with irregular dark stripes; Point Quobba northwards; Indo-E. Pacific; to 65 cm. ★★★

803 SCHLEGEL'S PARROTFISH
Scarus schlegeli (Bleeker)
Inhabits coral reefs; male distinguished by 2 yellow patches below dorsal fin, the rear one forming a narrow bar between the dorsal and anal fin, female by overall dark colour and 4-5 narrow pale bars (often faint, sometimes absent) on side; Abrolhos northwards; mainly W. and C. Pacific; to 35 cm. ★★★

804 GREEN-FINNED PARROTFISH
Scarus sordidus Forsskål
Inhabits coral reefs; male distinguished by overall green colour with pale tail base and bluish cheeks, throat and breast, female by white tail with round black spot at base; Rottnest Island northwards, but rare south of Abrolhos; Indo-C. Pacific; to 45 cm. ★★★

805 SURF PARROTFISH
Scarus rivulatus Valenciennes
Inhabits coral reefs; male distinguished by orange patch on cheek, wavy lines on snout, and light yellow-green colour of pectoral fins, females are variable, but generally very pale; Ningaloo Reef northwards; mainly W. Pacific; to 45 cm. ★★★

BOTTOM PREDATORS
(See Plate 54)

The fishes illustrated on Plate 54 are typically bottom dwellers which occur in sandy areas. The Grubfishes (806-813) belong to the family Mugiloididae. The group contains an estimated 45 species which mainly occur in the Indo-Pacific region. They feed on small fishes, crustaceans, shellfish, and worms. Most are relatively small (under 25 cm).

Jawfishes of the family Opistognathidae 814-818 occupy elaborate burrows lined with stones. The Leopard Jawfish (818) has been known to attack the feet of waders.

Stargazers of the family Uranoscopidae (820-824) are scaleless, sand-dwelling fishes with a large 'trap-door' mouth and eyes that face upwards. They are found in tropical and temperate seas at depths ranging from about 15 m to deeper sections of the continental shelf. Stargazers sometimes bury themselves leaving only the mouth and eyes exposed. Their diet consists mainly of fishes.

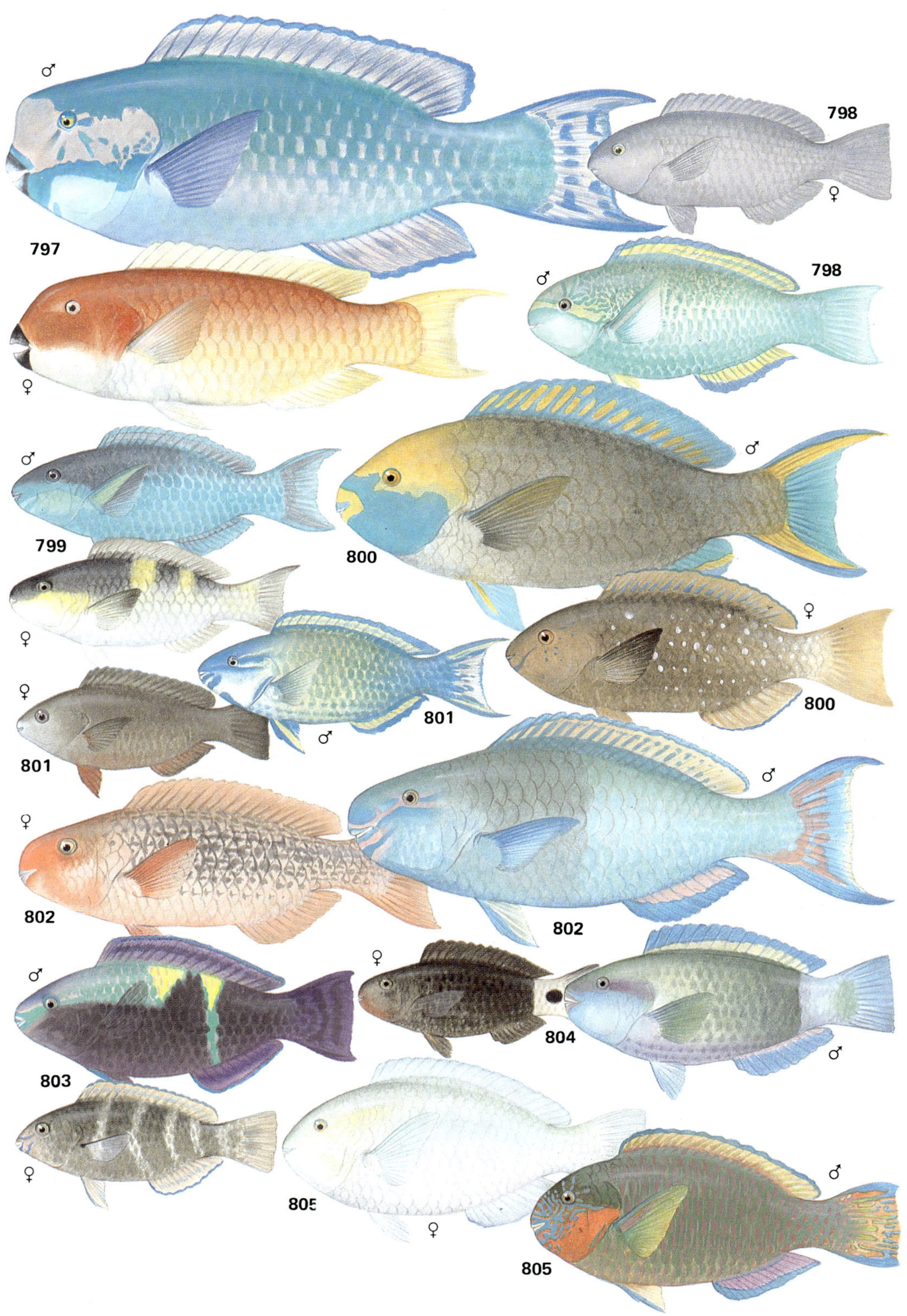

806 BLUE-NOSED GRUBFISH
Parapercis alboguttata (Günther)
Inhabits trawling grounds; distinguished by diffuse dark bars or blotches on side and bluish snout; Shark Bay northwards; N. Indian Ocean and W. Pacific; to 22 cm.

807 DOUBLESPOT GRUBFISH
Parapercis diplospilus Gomon
Inhabits trawling grounds; distinguished by diffuse dark spots or blotches on side and 2 spots on tail base; Exmouth Gulf northwards; Indo-Australian Archipelago; to 9 cm.

808 SPOTHEAD GRUBFISH
Parapercis clathrata Ogilby
Inhabits coral reefs; distinguished by row of enlarged spots along lower side, these usually connected by a thin stripe, also frequently with pale-edged black spot above gill cover; Dirk Hartog Island northwards; Indo-C. Pacific; to 17 cm.

809 NARROW BARRED GRUBFISH
Parapercis macrophthalma (Pietschmann)
Inhabits trawling grounds; distinguished by 5 narrow bars on side and dark spot on upper tail base; Broome northwards; mainly W. Pacific; to 14 cm.

810 ROSY GRUBFISH
Parapercis gushikeni Yoshino
Inhabits trawling grounds; distinguished by narrow lines on upper side and spotted tail with pointed upper edge; Port Hedland northwards; mainly W. Pacific; to 30 cm.

811 BANDED GRUBFISH
Parapercis mimaseana (Kamohara)
Inhabits trawling grounds; distinguished by black spot at front of dorsal fin, wavy stripes on upper side, and narrow bars on tail; Dampier northwards; mainly W. Pacific; to 20 cm.

812 RED-BANDED GRUBFISH
Parapercis multiplacata Randall
Inhabits offshore coral reefs; distinguished by black spiny dorsal fin and series of red bars on side; Ningaloo Reef northwards; mainly W. Paciific; to 10 cm.

813 RED-BARRED GRUBFISH
Parapercis nebulosa (Quoy & Gaimard)
Inhabits trawling grounds; similar to 812, but cross bars wider and more diffuse, also lacks row of dark spots along lower side; Abrolhos northwards; mainly Indian Ocean; to 20 cm.

814 ABROLHOS JAWFISH
Opistognathus sp.
Inhabits rubble and sand bottoms below 20 m depth; distinguished by small size and irregular dark longitudinal band on sides; known thus far only from the Abrolhos; to 10 cm.

815 DARWIN JAWFISH
Opistognathus darwiniensis Macleay
Inhabits shallow reefs, usually in sandy or rubble areas; distinguished by yellowish colour of fins, dense spotting on head and sides, and prominent banding on dorsal, anal and tail fins; Ningaloo Reef northwards; N. Australia only; to 50 cm.

816 BLOTCHED JAWFISH
Opistognathus latitabundus (Whitley)
Inhabits coastal waters, usually on rubble or soft bottoms; distinguished by large dark blotches on back and base of dorsal fin; also known as Spotted pug; Broome northwards; N. Australia only; to 30 cm.

817 BLACK JAWFISH
Opistognathus inornatus Ramsay & Ogilby
Inhabits coastal waters, usually on rubble or soft bottoms; distinguished by overall dark colour without markings; Exmouth Gulf northwards; N. Australia only; to 55 cm.

818 LEOPARD JAWFISH
Opistognathus reticulatus McKay
Inhabits coastal waters, usually on rubble or soft bottoms; distinguished by large black spots; has bitten people wading in shallow water, but usually harmless; Exmouth Gulf northwards; N. Australia only; to 50 cm.

819 BANDED STARGAZER
Ichthyscopus fasciatus Haysom
Inhabits trawling grounds; distinguished by 5 dark bars on upper side; Broome northwards; N. Australia only; to 25 cm.

820 DOUBLE-BANDED STARGAZER
Ichthyscopus insperatus Mees
Inhabits trawling grounds; distinguished by narrow double-bars on back and barred tail; Shark Bay northwards; N. Australia only; to 30 cm.

821 MARBLED STARGAZER
Uranoscopus bicinctus Temminck & Schlegel
Inhabits trawling grounds; distinguished by white patches on back and diffuse broad bar below each dorsal fin, has stout spine behind upper edge of gill cover; Dampier northwards; mainly W. Pacific; to 20 cm.

822 YELLOWTAIL STARGAZER
Uranoscopus cognatus Cantor
Inhabits trawling grounds; distinguished by overall brown colour except for black spot on first dorsal fin and lighter tail, has stout spine behind upper edge of gill cover; Shark Bay northwards; Indo-Australian Archipelago; to 22 cm.

823 KAI STARGAZER
Uranoscopus kaianus Günther
Inhabits trawling grounds; similar to 822, but has variegated pattern on upper half of body; Port Hedland northwards; Indo-Australian Archipelago; to 22 cm.

824 WHITE-SPOTTED STARGAZER
Uranoscopus sp.
Inhabits trawling grounds; distinguished by small white spots on upper part of head and body, has stout spines behind upper edge of gill cover; Derby northwards, possibly N. Australia only; to 15 cm.

825 SLENDER SABRETOOTH BLENNY
Aspidontus dussumieri (Valenciennes)
Inhabits coral reefs and southern rocky reefs; similar to 826, but has yellow fins; Jurien Bay northwards; Indo-C. Pacific; to 14 cm.

826 FALSE CLEANERFISH
Aspidontus taeniatus Quoy & Gaimard
Inhabits coral reefs; nearly identical in colour to the Cleanerfish (765), but has more pointed snout, longer dorsal fin base, and enlarged fangs at back of lower jaw; uses its "disguise" to attack other fishes taking bites of skin and scales; Shark Bay northwards; Indo-C. Pacific; to 12 cm.

827 BROWN CORAL BLENNY
Atrosalarias fuscus holomelas (Günther)
Inhabits coral reefs; distinguished by blackish to brown body and relatively tall dorsal and anal fins with abruptly pale tail; an entirely yellow colour phase is sometimes seen; Abrolhos northwards; Indo-C. Pacific; to 10 cm.

828 MANY-SPOTTED BLENNY
Laiphognathus multimaculatus Smith
Inhabits coral reefs; distinguished by numerous small red spots and 1-2 rows of white spots just below base of dorsal fin; Shark Bay northwards; W. Pacific and E. Indian Ocean; to 5.5 cm.

829 MIMIC BLENNY
Mimoblennius atrocinctus (Regan)
Inhabits coral reefs; distinguished by faint bars and row of double-spots along midside, underside of head is dusky to black; Dampier Archipelago northwards; W. Pacific and E. Indian Ocean; to 4 cm.

830 TALBOT'S BLENNY
Stanulus talboti Springer
Inhabits coral reefs; distinguished by small white spots on cheek and lower part of head, larger white spots or blotches on body, and dark irregular blotches above pectoral fin; Dirk Hartog Island northwards; Indo-W. Pacific; to 7 cm.

831 LEOPARD BLENNY
Exallias brevis (Kner)
Inhabits coral reefs; distinguished by dense network; of leopard-like spotting; Ningaloo Reef northwards; Indo-C. Pacific; to 12 cm.

832 SHORT-HEADED SABRETOOTH BLENNY
Petroscirtes breviceps Valenciennes
Inhabits weed-sand areas in the vicinity of reefs; distinguished by mottled colour, usually with dark stripe or broken band along middle of side; also known as Weed blenny; Cockburn Sound northwards; Indo-W. Pacific; to 15 cm.

833 HIGH-FINNED BLENNY
Petroscirtes mitratus Rüppell
Inhabits weed-sand areas in the vicinity of reefs; distinguished by tall "mast" at front of dorsal fin, juveniles similar to 832, but slightly deeper-bodied; Rottnest Island northwards; Indo-W. Pacific; to 7 cm.

834 BLACK-BANDED BLENNY
Meiacanthus grammistes (Valenciennes)
Inhabits sand-weed areas and coral reefs; distinguished by black stripes that become broken into spots on rear of body and tail; Abrolhos northwards; mainly W. Pacific; to 10 cm.

835 GERMAIN'S BLENNY
Omobranchus germaini (Sauvage)
Inhabits shallow reefs, usually in crevices just below the level of low tide; distinguished by "ear" spot and series of narrow white to blue lines on side; Geographe Bay northwards; mainly W. Pacific; to 8 cm.

836 ROTUND BLENNY
Omobranchus ferox (Herre)
Inhabits mangrove bays or estuaries; distinguished by semi-transparent body and white streak behind eye; Monte Bello Islands northwards; Indian Ocean and Indo-Australian Archipelago; to 7 cm.

837 ROUND-HEADED BLENNY
Omobranchus lineolatus (Kner)
Inhabits mangrove bays and estuaries; distinguished by "ear" spot, pronounced dark bars on head, and fainter irregular bars on side; Broome northwards; N. Australia and Melanesia; to 9 cm.

838 MUZZLED BLENNY
Omobranchus punctatus (Valenciennes)
Inhabits inshore reefs and estuaries; distinguished by vertical bars on head and broken thin stripes on side; Abrolhos northwards; W. Pacific and E. Indian Ocean; to 9 cm.

839 YELLOW SABRETOOTH BLENNY
Plagiotremus tapeinosoma (Klunzinger)
Inhabits coral and rocky reefs; distinguished by broad black stripe which becomes broken (or has wavy margin) towards rear of body, and broad dark margins on dorsal and anal fins; this fish and 840 have a pair of enlarged fangs and sometimes bite divers; Rottnest Island northwards; Indo-E. Pacific; to 13 cm.

840 BLUE-LINED SABRETOOTH BLENNY
Plagiotremus rhinorhynchus (Bleeker)
Inhabits coral and rocky reefs; distinguished by narrower blue stripes on each side of broad brownish to orange stripe, and has pale fins; Walpole northwards, but relatively rare south of the Abrolhos; Indo-C. Pacific; to 12 cm.

841 HAIR-TAIL BLENNY
Xiphasia setifer Swainson
Inhabits sandy areas in the vicinity of coral reefs; distinguished by extremely elongate, tapering body; Exmouth Gulf northwards; Indo-W. Pacific; to 56 cm.

FANGED FISHES

Blennies of the family Blennidae (Plates 55-56) are small, elongate, scaleless fishes that are common on shallow reefs, mainly in tropical seas. Most of the estimated 300 species occurring worldwide are under 10-15 cm. They feed mainly on small invertebrates, algae, and bottom detritus. Several of the species shown on Plate 55 (825-827, 832-834, 839-841) are commonly referred to as saber-toothed blennies. The name refers to the large teeth of the lower jaw which appear to be used mainly for defense. Scale and fin predation is a characteristic feeding habit of *Aspidontus* (825-826) and *Plagiotremus* (839-840). The False Cleanerfish mimics the Cleaner Wrasse (765).

842 DUSKY BLENNY
Cirripectes filamentosus Alleyne & Macleay
Inhabits inshore reef crevices, frequently where there is wave action; distinguished by fringe of tentacles on neck, overall dusky colour and elevated fin rays at front of dorsal fin; the Black blenny, an undescribed species (not shown), is similar, but usually has small black spots on the head and front part of body and has a lower dorsal fin, it ranges from Cape Leeuwin to Dampier Archipelago, but is less common than the Dusky blenny in the northern part of its range; Shark Bay northwards; Australia only; to 10 cm.

843 SPOTTED-CHIN BLENNY
Cirripectes sebae
Inhabits inshore reef crevices exposed to surge; distinguished by fringe of tentacles on neck, pale spots on chin, and series of narrow dark bars on side (sometimes barely visible); Abrolhos northwards; Indo-C. Pacific; to 13 cm.

844 BICOLOR BLENNY
Ecsenius bicolor (Day)
Inhabits coral reef crevices; 2 colour varieties are common: one that is bright orange posteriorly and another that is overall dark brown; Shark Bay northwards; Indo-W. Pacific; to 8 cm.

845 LINED BLENNY
Ecsenius lineatus Klausewitz
Inhabits coral reefs; distinguished by black stripe on upper side with brown area above and abruptly whitish on lower half; Ningaloo northwards; W. Pacific and E. Indian Ocean; to 7 cm.

846 OCULAR BLENNY
Ecsenius oculus Springer
Inhabits coral reef crevices; distinguished by series of pale-edged black spots connected by pale lines on side; Dirk Hartog Island northwards; mainly W. Pacific; to 6.5 cm.

847 CORAL BLENNY
Ecsenius yaeyamensis (Aoyagi)
Inhabits coral reef crevices; distinguished by dark stripe behind eye, pale spots on side, and pair of thin lines across base of pectoral fin; Ningaloo Reef northwards; mainly W. Pacific; to 6.5 cm.

848 TWINSPOTS BLENNY
Entomacrodus thalassinus (Jordan & Seale)
Inhabits shallow coral reefs; distinguished by dark streak behind eye, row of large spots arranged in pairs along middle of side and spotted dorsal and tail fins; Monte Bello Islands northwards; Indo-W. Pacific; to 6.5 cm.

849 WAVY-LINED BLENNY
Entomacrodus decussatus (Bleeker)
Inhabits shallow reefs exposed to wave action; distinguished by wavy lines and irregular light and dark blotches on sides; Ningaloo Reef northwards; Indo-C. Pacific; to 18 cm.

850 STRIATED BLENNY
Entomacrodus striatus (Quoy & Gaimard)
Inhabits shallow reefs exposed to wave action; distinguished by clusters of dark spots on back superimposed on 4-5 broad diffuse dark bars; Rottnest Island northwards; Indo-C. Pacific; to 10 cm.

851 RED-SPOTTED BLENNY
Istiblennius chrysospilos Bleeker
Inhabits shallow reefs and tide pools exposed to wave action; distinguished by small red spots on head and body, and series of dark vertical streaks arranged in pairs on side; Abrolhos northwards; mainly W. Pacific; to 13 cm.

852 RIPPLED BLENNY
Istiblennius edentululus (Bloch & Schneider)
Inhabits shallow reefs and tide pools exposed to wave action; male has skin flap on top of head, and dark bars (usually in pairs) and pale streaks on side; female has lighter bars on sides, spots or lines on dorsal and anal fins, and often with numerous red to brown spots on rear part of body; Dirk Hartog Island northwards; Indo-C. Pacific; to 13 cm.

853 SPOTTED BLENNY
Istiblennius meleagris (Valenciennes)
Inhabits shallow reefs and tide pools exposed to wave action; male with low skin flap on top of head and rows of pale spots on side, female lacks skin flap and has faint forward-slanting bars and scattered pale spots on side; Port Denison northwards; Indo-W. Pacific; to 15 cm.

854 TIDEPOOL BLENNY
Istiblennius lineatus (Valenciennes)
Inhabits shallow reefs and tide pools exposed to wave action; distinguished by skin flap on head and general pale colouration with narrow longitudinal lines on side; Ningaloo Reef northwards; Indo-C. Pacific; to 20 cm.

855 BLUE-STREAKED BLENNY
Istiblennius periophthalmus (Valenciennes)
Inhabits shallow reef and tide pools exposed to wave action; distinguished by H-shaped bars on side and double row of silvery-blue streaks or spots on middle of side; Dampier Archipelago northwards; Indo-W. Pacific; to 14 cm.

856 BANDED BLENNY
Salarias fasciatus (Bloch)
Inhabits inshore coral reefs, often in weeds; distinguished by light and dark bars with overlay of narrow dark lines; Ningaloo Reef northwards; Indo-W. Pacific; to 13 cm.

857 STARRY BLENNY
Salarias sp.
Inhabits sand-weed areas on rocky outcrops; distinguished by frilly tentacles above eye and numerous small white spots; possibly an undescribed species; Ningaloo Reef northwards; Indo-Australian Archipelago; to 9 cm.

858 SPALDING'S BLENNY
Salarias spaldingi Macleay
Inhabits coral reefs; distinguished by narrow dark bars on side and 'pepper'-spotting on upper side; Exmouth Gulf northwards; N. Australia only; to 9 cm.

> ### TIDEPOOL BLENNIES
>
> Most of the blennies shown on this plate, particularly 848-858, are common inhabitants of shallow reef flats, the intertidal zone and splash pools along rocky shores. They are one of the most frequently observed groups of fishes encountered by beachcombers. A few of the species are commonly called rock-skippers because if disturbed they use their body musculature and stout pelvic fins to skip over rocks, out of water, as they leap from pool to pool.

859 BLACK-THROATED THREEFIN
Helcogramma decurrens McCulloch & Waite
Inhabits inshore reefs; male distinguished by black on lower half of head and body, female is variably coloured, either red, green, or brown; entire southern coast of W.A. north to Point Quobba; S.A. and W.A. only; to 7 cm.

860 NEON THREEFIN
Helcogramma sp.
Inhabits coral reef crevices; distinguished by reddish colour with 3 narrow golden stripes on side; Ningaloo Reef northwards; possibly N. Australia only; to 3.5 cm.

861 GREEN THREEFIN
Enneapterygius sp.
Inhabits inshore reefs; genus similar to *Helcogramma* (859 and 860), but has straight or only slightly curved lateral line that ends on middle of side, but then re-continues to tail base as series of notched scales (versus curved downwards behind pectoral fin and terminating on middle of side without notched scales); there are a number of apparently undescribed members of this genus in northwestern waters; this species distinguished by blotchy green colour with pale saddles on back; Abrolhos northwards; entire distribution unknown; to 3.5 cm.

862 GOODLAD'S STINKFISH
Callionymus goodladi (Whitley)
Inhabits trawling grounds, generally on sand bottoms; distinguished by overall pale colouration with brown mottling on back and spotted tail, male has 2 filamentous spines at front of first dorsal fin and female (not shown) has one filament; Esperance to Exmouth Gulf and offshore reefs of North West Shelf; W.A. only; to 22 cm.

863 QUEENSLAND STINKFISH
Callionymus moretonensis Johnson
Inhabits trawling grounds, generally on sand bottoms; similar to 864 and 865, but has shorter tail and stout spine on lower edge of cheek has 2 large teeth on its upper surface (versus many small teeth); Joseph Bonaparte Gulf; N. Australia and Melanesia; to 12 cm.

864 AUSTRALIAN STINKFISH
Callionymus margaretae australis Fricke
Inhabits trawling grounds, generally on sand bottoms; similar to 863, but has 5-7 small teeth on upper surface of cheek spine (versus 2 large teeth); also resembles 865, but has only outer tips of anal fin rays blackish (versus outer half of entire fin); Shark Bay northwards; N.W. Australia only, but another subspecies *C. margaretae margaretae* widespread in N. Indian Ocean; to 18 cm.

865 JAPANESE STINKFISH
Callionymus japonicus Houttyn
Inhabits trawling grounds, generally on sand bottoms; similar to 863 and 864 but has wider black margin on anal fin and tail often wider and more elongate; Dampier northwards; mainly W. Pacific; to 30 cm.

866 GROSS'S STINKFISH
Callionymus grossi Ogilby
Inhabits trawling grounds, generally on sand bottoms; similar to 862, but has larger, differently shaped dorsal fin; Abrolhos northwards; N. Australia only; to 20 cm.

867 ROSY DRAGONET
Synchiropus altevelis (Temminck & Schlegel)
Inhabits trawling grounds in deep (70-600 m) water; distinguished by overall reddish or pink colour and relatively short first dorsal fin (sometimes with filament); Broome northwards; mainly W. and C. Pacific; to 20 cm.

868 MORRISON'S DRAGONET
Synchiropus morrisoni Schultz
Inhabits coral reefs and rocky areas usually below 25 m depth; male distinguished by tall flat-topped first dorsal fin with dark vertical streaks and broad dark submarginal band on anal fin, female by large dark blotch covering pectoral fin base and diagonal dark streaks on anal fin; North West Cape northwards; mainly W. Pacific; to 6.5 cm.

869 HIGH-FINNED DRAGONET
Synchiropus rameus McCulloch
Inhabits trawling grounds, generally on sand or rubble bottoms; distinguished by huge sail-like dorsal fin and blue-spotted anal fin; Shark Bay northwards; N. Australia and New Caledonia; to 15 cm.

870 WESTERN DRAGONET
Synchiropus picturatus occidentalis Fricke
Inhabits coral reefs; distinguished by robust shape and ornate pattern of ocellated spots and blotches; known only from Dampier Archipelago, another subspecies *S. picuratus picturatus* occurs in the Indo-Malayan area; to 5.5 cm.

871 NORTHERN DRAGONET
Diplogrammus xenicus (Jordan & Thompson)
Inhabits sand-rubble areas near reefs; male distinguished by relatively short filament on first dorsal fin, blue spots and lines on head, dusky anal fin, and fan-like, dark-edged pelvic fins, female differs from male in size, shape, and colour of first doral and pelvic fins, also the anal fin is clear instead of dusky; Abrolhos northwards; N.W. Australia and Okinawa to Japan; to 7 cm.

872 FINGERED DRAGONET
Dactylopus dactylopus (Valenciennes)
Inhabits sand-weed bottoms; distinguished by detached first ray on large fan-like pelvic fin; Cockburn Sound northwards; W. Pacific and E. Indian Ocean; to 30 cm.

TRIPLEFINS AND STINKFISHES

The Threefins of the family Tripterygiidae (859-861) as their name suggests are characterised by three separate dorsal fins. They are mainly tiny fishes with perhaps as many as 100 species occurring worldwide. The classification of the group is poorly documented and identification of most Australian species is therefore difficult. They are bottom dwellers that are usually found on the surface of rocks or corals.

Stinkfishes of the family Callionymidae (862-872) occur in tropical and temperate seas where they generally are found on sand or mud bottoms in the vicinity of reefs. Some species occur on the continental shelf and slope to depths of about 900 m. Nearly all of the estimated 135 species are small in size (usually under 20 cm TL) and most are inhabitants of the Indo-Pacific region.

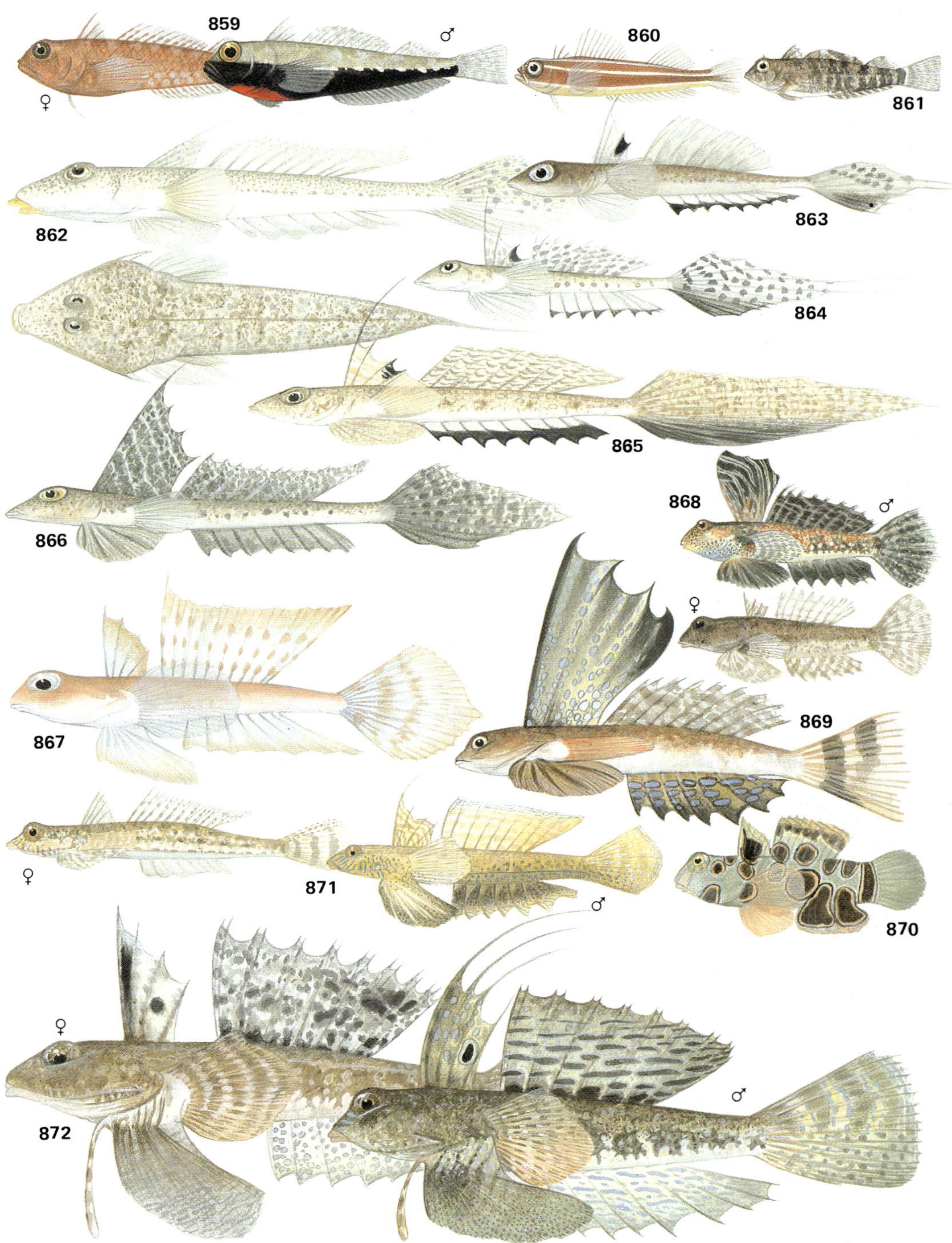

873 MANGROVE GOBY
Acentrogobius gracilis (Bleeker)
Inhabits mangrove estuaries; distinguished by overall pale colour with midlateral stripe composed of brown blotches and bright blue spots, also brown spots and blotches on back and fins; Dampier Archipelago northwards; Indo-Australian Archipelago; to 8 cm.

874 WHEELER'S SHRIMP GOBY
Amblyeleotris wheeleri (Polunin & Lubbock)
Inhabits sand bottoms near coral reefs, shares its burrow with an alpheid prawn; distinguished by combination of wine-red bars and small bluish spots; Ningaloo Reef northwards; Indo-W. Pacific; to 6 cm.

875 STARRY GOBY
Asterropteryx semipunctatus (Rüppell)
Inhabits rubble bottoms; distinguished by mottled green-brown colour with numerous small blue spots on head, body and fins; Abrolhos northwards; Indo-C. Pacific; to 5 cm.

876 BANDED GOBY
Amblygobius phalaena (Valenciennes)
Inhabits sand-weed areas; distinguished by stripes that are interrupted by narrow dark bars on side, a large pale-edged dark spot on first dorsal fin, and smaller dark spot on upper part of tail; also known as White spotted goby and Barred goby; Cockburn Sound northwards; Indo-W. Pacific; to 15 cm.

877 COCOS GOBY
Bathygobius cocosensis Bleeker
Inhabits shallow beach-rock reefs and tide pools distinguished by diffuse dark saddles and row of elongated dark spots slightly below middle of side, also scales on top of head extend only to rear margin of cheek (preopercle); Point Quobba northwards; Indo-W. Pacific; to 8 cm.

878 COMMON GOBY
Bathygobius fuscus (Rüppell)
Inhabits shallow beach-rock reefs and tide pools; distinguished by blunt snout, bulbous cheeks, and often with large irregular shaped blotches and small pale spots on head and side, scales on top of head extend almost to rear of eyes; Abrolhos northwards; Indo-C. Pacific; to 12 cm.

879 WHISKERED GOBY
Callogobius sp.
Inhabits rubble areas and coral reef crevices; distinguished by rows of raised papillae ("whiskers") on head, easily shed scales, and blotchy colour pattern; Abrolhos northwards; possibly Australia only; to 6 cm.

880 DOUBLE-BARRED GOBY
Callogobius sclateri (Steindachner)
Inhabits coral reefs; distinguished by pair of dark bars, one below each dorsal fin, and rows of raised papillae ("whiskers") on head; Ningaloo Reef northwards; Indo-C. Pacific; to 5 cm.

881 GREEN SHRIMP GOBY
Cryptocentrus caeruleomaculatus (Herre)
Inhabits sand-rubble areas near coral reefs, shares its burrow with an alpheid prawn; distinguished by broad green-brown bars with narrower pale areas between them, also blackish spot in middle of each dark bar; Dampier Archipelago northwards; Indo-Australian Archipelago; to 5 cm.

882 PINK SHRIMP GOBY
Cryptocentrus obliquus (Herre)
Inhabits sandy areas near coral reefs, shares its burrow with an alpheid shrimp; distinguished by pink or red spots and stripes on head and dorsal fins, and oblique dark bars on body, also small white spots on head and side; Dampier Archipelago northwards; Indo-Australian Archipelago; to 10 cm.

883 SPOTTED SHRIMP GOBY
Ctenogobiops pomastictus Lubbock & Polunin
Inhabits sandy areas near coral reefs, shares its burrow with an alpheid shrimp; distinguished by brown spots on head and body, and white blotch on lower edge of pectoral fins; Ningaloo Reef northwards; E. Indian Ocean and W. Pacific; to 6 cm.

884 MUD GOBY
Drombus sp.
Inhabits coastal waters, including estuaries; distinguished by abruptly pale back and small white spots on head and side; Broome northwards; N. Australia; to 5 cm.

885 RED CORAL GOBY
Eviota sp.
Inhabits coral reefs; distinguished by tiny size, semi-transparent appearance, red spots on head, black spot above gill cover, diffuse dark bars above anal fin, and dark spot on middle of tail base; Abrolhos northwards; possibly Australia only; to 2 cm.

886 SMITH'S CORAL GOBY
Eviota infulata (Smith)
Inhabits coral reefs; distinguished by tiny size, filamentous first and second dorsal spines (males) and black spot or blotch above pectoral fin base; Abrolhos northwards; Indo-C. Pacific; to 2 cm.

887 TWOSPOT GOBY
Fusigobius duospilus Hoese & Reader
Inhabits sand bottoms adjacent to coral reefs; distinguished by semi-transparent appearance, 2 dark blotches on first dorsal fin, small brown spots on head and body, and black spot at middle of tail base; Abrolhos northwards, Indo-C. Pacific; to 6 cm.

888 ESTUARY GOBY
Glossogobius biocellatus (Valenciennes)
Inhabits brackish estuaries and tidal creeks; distinguished by dark blotches or bars on lower lobe of tail, also anal and pelvic fins and lower part of head frequently dark; Onslow northwards; mainly W. Pacific; to 10 cm.

889 CIRCUMSPECT GOBY
Glossogobius circumspectus (Macleay)
Inhabits brackish estuaries and tidal creeks; similar to 888, but has blunter snout, is generally lighter in colour, and lacks darkened areas on pelvic, anal, and tail fins (except narrow dark bands across tail); Broome northwards; Indo-Australian Archipelago; to 12 cm.

890 YELLOWSPOT GOBY
Gnatholepis inconsequens Whitley
Inhabits sand bottoms adjacent to coral reefs; distinguished by narrow dark stripes on side, thin bar below eye, and small orange spot above pectoral fin base; Rottnest Island northwards; Indo-Australian Archipelago; to 7 cm.

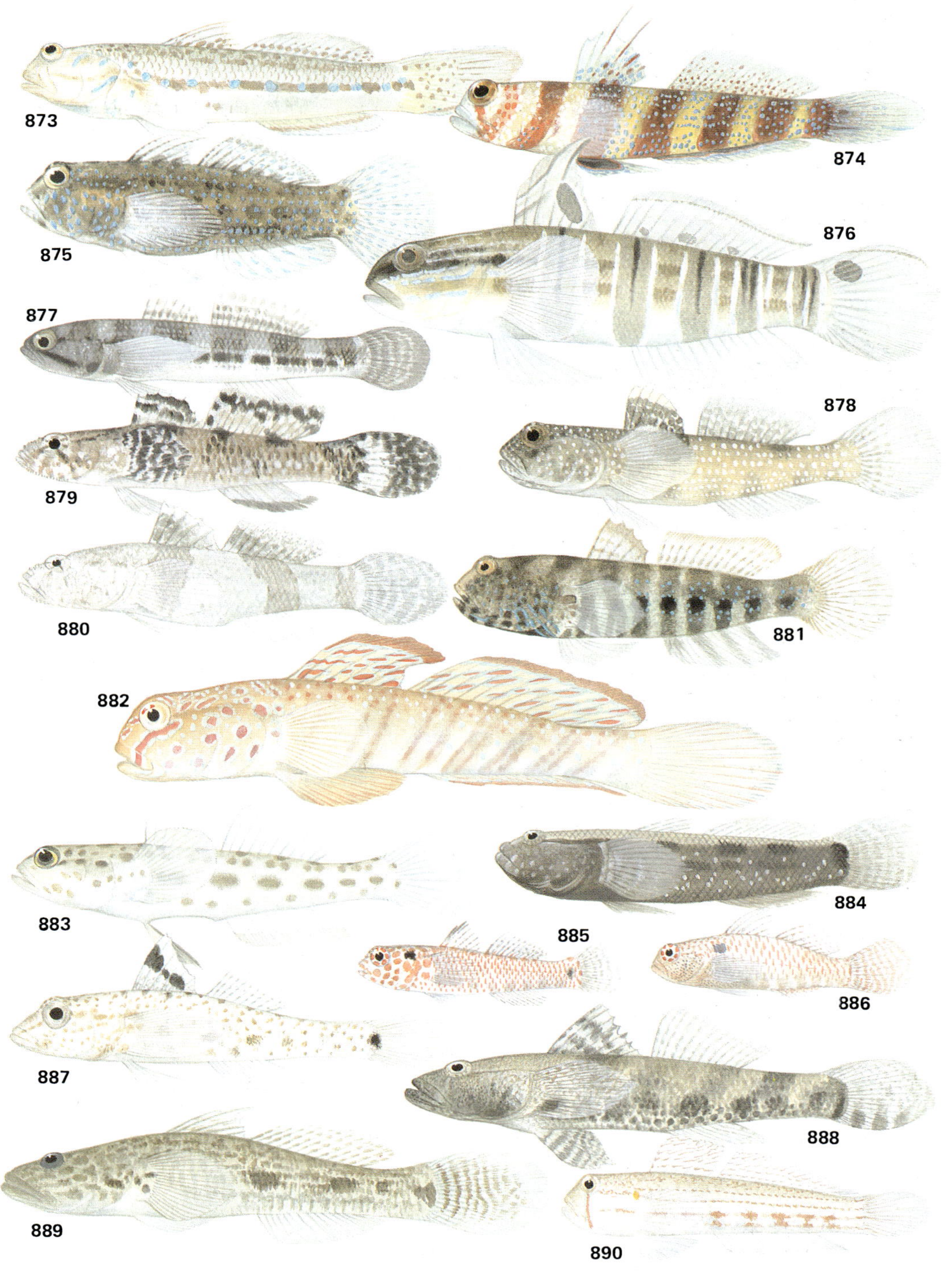

891 BROAD-BARRED MAORI GOBY
Gobiodon histrio (Valenciennes)
Inhabits coral reefs, wedging amongst branching corals; distinguished by large rounded head, and red-brown bars and stripes; Ningaloo Reef northwards; W. and C. Pacific; to 6 cm.

892 FIVE-BAR CORALGOBY
Gobiodon quinquestrigatus Bleeker
Inhabits coral reefs, wedging amongst branching corals; similar shape to 891, but overall dark with narrow blue lines on head; Dirk Hartog Island; Indo-C. Pacific; to 4 cm.

893 DECORATED GOBY
Istigobius decoratus (Herre)
Inhabits sand bottoms in the vicinity of coral reefs; similar to 894, but lacks black spot on rear part of first dorsal fin and few (none or less than 5) dark spots on top of head between eyes and first dorsal fin; Ningaloo Reef northwards; Indo-C. Pacific; to 12 cm.

894 BLACK-SPOTTED GOBY
Istigobius nigroocellatus (Günther)
Inhabits sand or silt bottoms in turbid water near inshore reefs; similar to 893, but has black spot at rear of first dorsal fin, and numerous (about 10-15) small black spots on top of head between eyes and first dorsal fin, and found in different habitat (silty turbid conditions versus clean coralline sand); Shark Bay northwards; mainly W. Pacific; to 7 cm.

895 SPECTACLED GOBY
Istigobius perspicillatus (Herre)
Inhabits sand-rubble areas near inshore reefs; distinguished by reddish colour, 12 rays in second dorsal fin, 12-16 scales in front of first dorsal fin along midline of head and narrow dark line connecting eyes over top of head; E. Indian Ocean and Indo-Australian Archipelago; to 11 cm.

896 ORNATE GOBY
Istigobius ornatus (Rüppell)
Inhabits mangrove areas and shallow rubble reefs; distinguished by 3-4 free (filamentous) upper pectoral rays and enlarged canine tooth on each side of lower jaw, colour variable: ornate pattern as shown or may be drab and similar to 893 or 894; Dampier Archipelago northwards; Indo-C. Pacific; to 10 cm.

897 SCHOOLING GOBY
Parioglossus formosus (Smith)
Inhabits mangrove areas, tidal creeks, and inshore reefs, occurring in schools; distinguished by prominent black stripe along lower side and extending on to tail; Dampier Archipelago northwards; mainly W. Pacific; to 3.5 cm.

898 GIRDLED GOBY
Priolepis cinctus (Regan)
Inhabits coral reef crevices and caves; distinguished by series of dark-edged brown bars with narrower pale bars between; Abrolhos northwards; Indo-C. Pacific; to 6 cm.

899 HEAD-BARRED GOBY
Priolepis semidoliatus (Valenciennes)
Inhabits coral reef crevices and caves; distinguished by narrow blue bars on head and elongate filament on first dorsal fin; Fremantle northwards; Indo-W. Pacific; to 3 cm.

900 REDHEAD GOBY
Paragobiodon echinocephalus (Rüppell)
Inhabits coral reefs, found amongst branches of *Stylophora* coral; distinguished by red or pinkish head and black body, also has tiny bristles covering head; Dampier Archipelago northwards; Indo-W. Pacific; to 3.5 cm.

901 BLUE GOBY
Ptereleotris hanae (Jordan & Snyder)
Inhabits sand bottoms near coral reefs; distinguished by slender shape, pale blue colour, and elongate filaments on first dorsal fin and tail; Dampier Archipelago northwards; mainly W. Pacific; to 10 cm.

902 RED AND WHITE GOBY
Trimma sp.
Inhabits coral reef crevices and caves; distinguished by tiny size, irregular red and white blotches on head and body, and red spots on fins; Ningaloo Reef northwards; entire distribution unknown; to 2.5 cm.

903 ORANGE-SPOTTED GOBY
Trimma okinawae (Aoyagi)
Inhabits coral reef crevices and caves; distinguished by dense network of red-orange spots on body and fins; Shark Bay northwards; E. Indian Ocean and W. Pacific; to 4 cm.

904 HOESE'S GOBY
Silhouettea hoesei Larson & Miller
Inhabits sand bottoms close to shore; distinguished by eyes on dorsal profile of head and overall pale colour with brown speckling; Ningaloo Reef northwards; N. Australia only; to 3.5 cm.

905 PATCHWORK GOBY
Pipidonia bravoi Herre
Inhabits inshore reefs; distinguished by flattened head, rows of raised papillae on head, short skin tentacles around mouth, and irregular dark bars and pale patches on body; Point Quobba northwards; mainly W. Pacific; to 4 cm.

906 STRIPED GOBY
Valenciennea muralis (Valenciennes)
Inhabits sand-rubble areas near coral reefs; distinguished by broad white band and series of narrower brown stripes along side and black spot at rear of first dorsal fin; Shark Bay northwards; E. Indian Ocean and W. Pacific; to 13 cm.

907 LONG-FINNED GOBY
Valenciennea longipinnis (Bennett)
Inhabits sand-rubble areas near coral reefs; distinguished by blue-edged saddles or bars on side with dark spot at lower part of each bar; Ningaloo Reef northwards; mainly W. Pacific; to 15 cm.

908 ORANGE-DASHED GOBY
Valenciennea puellaris (Tomiyama)
Inhabits sand-rubble areas near coral reefs; distinguished by orange longitudinal band from mouth to tail and orange spots on back; Abrolhos northwards; mainly W. Pacific; to 14 cm.

909 OCELLATED GOBY
Vanderhorstia ornatissimus (Smith)
Inhabits sand bottoms, lives in burrow with an alpheid shrimp; distinguished by long filament on first dorsal fin and blue-edged yellow spots on body; Abrolhos northwards; Indo-C. Pacific; to 8 cm.

910 SHADOW GOBY
Yongeichthys nebulosus (Forskål). Text on page 138.

891
892
893
894
895
896
897
898
899
900
901
902
903
904
905
906
907
908
909
910

911 MADURA GOBY
Apocryptodon madurensis Bleeker
Inhabits brackish mangrove estuaries; distinguished by overall whitish colour, scattered dark spots on head and body, and dark-edged pointed tail; Dampier northwards; Indo-W. Pacific; to 8 cm.

912 MUDSKIPPER
Periophthalmus argentiventralis Eggert
Inhabits brackish mangrove estuaries, often seen resting on muddy banks; several similar species in north-western waters, but this is one of the most common; distinguished by protruding eyes and sail-like dorsal fin; Carnarvon northwards; Indo-Australian Archipelago; to 27 cm.

913 BEARDED GOBY
Scartelaos histiophorus (Valenciennes)
Inhabits brackish mangrove estuaries; distinguished by a row of short barbels ("whiskers") along lower surface of head and antenna-like dorsal fin; Port Hedland northwards; mainly W. Pacific; to 15 cm.

914 GOGGLE-EYED GOBY
Boleophthalmus pectinirostris (Linnaeus)
Inhabits brackish mangrove estuaries; distinguished by huge first dorsal fin with filamentous edge; Broome northwards; mainly W. Pacific; to 12 cm.

915 SMALL-EYED SLEEPER
Prionobutis microps (Weber)
Inhabits brackish mangrove estuaries; distinguished by diagonal bands on head, striped fins, and robust shape; Broome northwards; Indo-Australian Archipelago; to 23 cm.

916 OLIVE FLATHEAD GUDGEON
Butis amboinensis (Bleeker)
Inhabits brackish mangrove estuaries; distinguished by flattened shovel-like snout; Broome northwards; S.E. Asia and Indo-Australian Archipelago; to 14 cm.

917 CHINESE GUDGEON
Bostrichthys sinensis (Lacepéde)
Inhabits coastal mudflats and estuaries; distinguished by pale-rimmed spot at upper tail base; Dampier Archipelago northwards; mainly W. Pacific and Andaman Sea; to 12 cm.

918 BLIND GOBY
Brachyamblyopus coecus (Weber)
Inhabits soft mud, usually buried below surface; distinguished by reddish-pink colour and lack of eyes; Broome northwards; Indo-Australian Archipelago; to 5 cm.

919 MOORISH IDOL
Zanclus cornutus (Linnaeus)
Inhabits coral reefs; distinguished by elongate snout, dorsal fin filament and conspicuous pattern; Point Quobba northwards; Indo-C. Pacific; to 24 cm.

920 SPOTTED SPINEFOOT
Siganus punctatus (Schneider)
Inhabits coral and rocky reefs; distinguished by numerous spots on head, body, and fins; Rottnest Island northwards; W. Pacific and E. Indian Ocean; to 40 cm. ★★★

921 BLACK SPINEFOOT
Siganus fuscescens (Houttyn)
Inhabits rock and weed areas, sometimes found in estuaries, occurs in schools; similar to 923, but body colour more uniform (generally lacks spotting) and edge of gill cover darkly outlined; Rottnest Island northwards; mainly W. Pacific; to 41 cm; 1.06 kg. ★★★

922 GOLDEN-LINED SPINEFOOT
Siganus lineatus (Linnaeus)
Inhabits coral reefs and mangrove estuaries; distinguished by yellow-orange lines and large spot below rear part of dorsal fin; Shark Bay northwards; W. Pacific and E. Indian Ocean; to 30 cm. ★★

923 SMUDGESPOT SPINEFOOT
Siganus canaliculatus (Park)
Inhabits sand-weed areas; similar to 921, but has more prominent spots on body and fins, frequently with dark blotch behind upper edge of gill cover; Point Quobba northwards; Indo-W. Pacific; to 20 cm. ★★★

924 THREESPOT SPINEFOOT
Siganus trispilos Woodland & Allen
Inhabits coral reefs, often found amongst branching *Acropora* coral; distinguished by 3 black blotches on upper side; Ningaloo Reef complex; W.A. only; to 23 cm.

925 DOUBLEBAR SPINEFOOT
Siganus virgatus (Valenciennes)
Inhabits coral reefs; distinguished by pair of diagonal dark bars on head and front of body; Dampier Archipelago northwards; W. Pacific and E. Indian Ocean; to 30 cm. ★★

PLATE 59
910 SHADOW GOBY
Yongeichthys nebulosus (Forsskål)
Inhabits sand or silt bottoms, sometimes in estuaries; distinguished by 3 or 4 large dark brown spots and smaller spots on side, dorsal fins and tail also spotted; Shark Bay northwards; Indo-W. Pacific; to 18 cm.

GOBIES AND RABBITFISHES

Gobies (Plates 58 and 59; and Plate 60, 911-914) are the largest family of marine fishes with more than 1000 species, including an extimated 300 in Australia. Only a few of the more common ones are included here. They are small (usually under 10 cm), bottom fishes that are found in most tropical and temperate seas. Some species live in association with other animals such as shrimps, sponges, gorgonians, corals, and sea urchins. Many dwell in sandy areas along the fringe of the reef and seek shelter in burrows. They feed on a wide variety of items including invertebrates, algae, and plankton.

Rabbitfishes of the family Siganidae (920-925) are plant feeders and sometimes form large schools as they roam over the reef. They are also sometimes called 'spine-feet' in reference to the unusual arrangement of two pelvic fin spines that are separated by three soft rays. Another peculiarity is the high (seven) number of anal fin spines. All of the spines of the dorsal, anal, and pelvic fins are grooved and contain venom glands. If handled carelessly they are capable of inflicting very painful wounds.

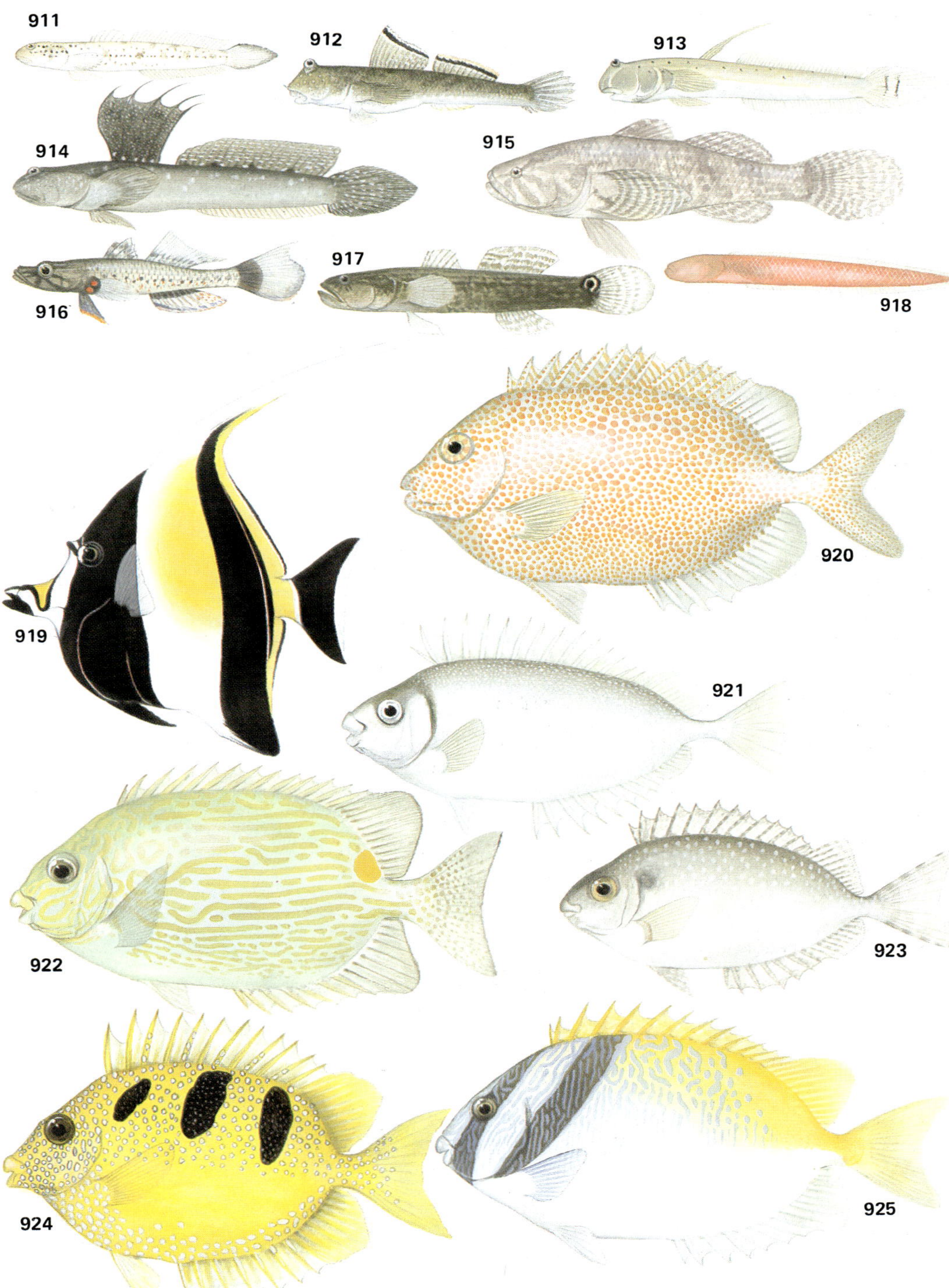

926 YELLOWMASK SURGEONFISH
Acanthurus mata Cuvier
Inhabits coral reef areas; distinguished by narrow stripes on side and yellow bands in front of eye; sharp spine on each side of tail base; formerly known as *A. bleekeri;* Ningaloo Reef northwards; Indo-C. Pacific; to 50 cm. ★★

927 WHITE-CHEEKED SURGEONFISH
Acanthurus nigricans (Linnaeus)
Inhabits coral reefs; distinguished by white patch below eye; sharp spine on each side of tail base; formerly known as *A. glaucoparieus;* Ningaloo Reef northwards; mainly Pacific Ocean to Central America and the Galapagos; to 21 cm.

928 ORNATE SURGEONFISH
Acanthurus dussumieri Valenciennes
Inhabits coral reefs; distinguished by yellowish fins and black outline around spine on base of tail; Ningaloo Reef northwards; Indo-W. Pacific; to 54 cm. ★★

929 BLUE-LINED SURGEONFISH
Acanthurus lineatus (Linnaeus)
Inhabits shallow coral reefs usually where there is some wave action; distinguished by blue stripes on side; sharp spine on each side of tail base; Ningaloo Reef northwards; Indo-C. Pacific; to 38 cm. ★★

930 RING-TAILED SURGEONFISH
Acanthurus grammoptilus Richardson
Inhabits coral and rocky reefs; distinguished by uniform dark colour and pale bar at base of tail; sharp spine on each side of tail base; Abrolhos northwards; Indo-Australian Archipelago; to 30 cm; .9 kg. ★★

931 ORANGE-SPOT SURGEONFISH
Acanthurus olivaceous Bloch & Schneider
Inhabits sand and rubble areas adjacent to coral reefs; distinguished by elliptical orange band behind eye, juveniles entirely yellow; sharp spine on each side of tail base; Ningaloo Reef northwards; Indo-C. Pacific; to 32 cm. ★★

932 DUSKY SURGEONFISH
Acanthurus nigrofuscus (Forsskål)
Inhabits coral and rocky reefs; distinguished by black spot at rear base of dorsal and anal fins; sharp spine on each side of tail base; Rottnest Island northwards; Indo-C. Pacific; to 20 cm. ★★

933 CONVICT SURGEONFISH
Acanthurus triostegus (Linnaeus)
Inhabits coral and rocky reefs; distinguished by bold black bars on side; sharp spine on each side of tail base; Rottnest Island northwards; Indo-E. Pacific; to 25 cm. ★★

934 BRISTLE-TOOTHED SURGEONFISH
Ctenochaetus strigosus (Bennett)
Inhabits coral reefs; distinguished by overall dark colour and flexible bristle-like teeth (versus fixed teeth); sharp spine on each side of tail base; Ningaloo Reef northwards; Indo-C. Pacific; to 18 cm. ★★

935 STRIPE-FACE UNICORNFISH
Naso lituratus (Bloch & Schneider)
Inhabits coral reefs; distinguished by dark stripe on snout and yellow patches at tail base; pair of sharp spines on each side of tail base; Ningaloo Reef northwards; Indo-C. Pacific; to 15 cm. ★★

936 LONGNOSED UNICORNFISH
Naso brevirostris (Valenciennes)
Inhabits coral reefs; distinguished by long spike on snout and short white tail; a pair of sharp spines on each side of tail base; Ningaloo Reef northwards; Indo-C. Pacific; to 50 cm. ★★

937 HUMPHEAD UNICORNFISH
Naso tuberosus Lacepède
Inhabits coral reefs; distinguished by bulbous snout and plain greyish colour; *Naso fageni* (Plate 70) is similar and more abundant, but lacks the forehead hump; pair of sharp spines on each side of tail base; Cape Cuvier northwards; Indo-W. Pacific; to 60 cm.

938 BROWN UNICORNFISH
Naso unicornis (Forsskål)
Inhabits coral reefs; distinguished by relatively short forehead spike, blue spots around pair of spines on each side of tail base, and elongate tail filaments; Ningaloo Reef northwards; Indo-C. Pacific; to 70 cm. ★★

938a BLUNT UNICORNFISH
Naso fageni Morrow
See Plate 70.

939 SAILFIN TANG
Zebrasoma veliferum (Bloch)
Inhabits coral reefs; distinguished by sail-like fins and bars on side; sharp spine on each side of tail base; Abrolhos northwards; Indo-C. Pacific; to 40 cm. ★★

940 BLUE-LINED TANG
Zebrasoma scopas (Valenciennes)
Inhabits coral reefs; distinguished by overall dark colour, protruding snout and white spine on each side of tail base; Abrolhos northwards; Indo-C. Pacific; to 20 cm. ★★

SURGEONFISHES

Surgeonfishes of the family Acanthuridae are widely distributed in tropical and subtropical seas. Most of the estimated 75 species occur in the Indo-west Pacific region. They are solitary in habit or form schools, sometimes containing more than one species and up to several hundred individuals. The largest genus, *Acanthurus* (926-933), contains mainly algal feeders that graze widely over their home reefs. However, the young often feed mainly on zooplankton and tiny benthic invertebrates. Most members of the genus *Naso* (935-938) and a few *Acanthurus* feed high above the bottom on zooplankton. Surgeonfishes derive their common name from the scalpel-like spine present on each side of the tail base which is used as a defensive weapon. In *Acanthurus, Ctenochaetus* (934), and *Zebrasoma* (939-940) there is a single movable spine, whereas *Naso* possesses a pair of fixed spines.

926
928
927
930
929
juv.
932
931
934
933
935
936
937
939
938
940
juv.

941 BLACK MARLIN
Makaira indica (Cuvier)
Inhabits oceanic waters, generally well offshore; distinguished by lack of cross bars and the rigid pectoral fin which cannot be folded against side of body; all West Australian seas; Indo-C. Pacific; to 500 cm. All tackle world record 707.61 kg, Australian record 199.580 kg. ★★★

942 INDO-PACIFIC BLUE MARLIN
Makaira mazara (Jordan & Snyder)
Inhabits oceanic waters, generally well offshore; distinguished from 941 by narrow bars on sides and non-rigid pectoral fin, and from 944 by its lower dorsal fin; Albany northwards; Indo-E. Pacific; to 500 cm. All tackle world record 498.95 kg, Australian record 275.0 kg. ★★★

943 SHORT BILL SPEARFISH
Tetrapturus angustirostris Tanaka
Inhabits oceanic waters, generally well offshore; distinguished by very short bill and slender shape; Cape Leeuwin northwards; Indo-E. Pacific; to 200 cm. ★★

944 STRIPED MARLIN
Tetrapturus audax (Philippi)
Inhabits oceanic waters, generally well offshore; similar to 942 but has taller dorsal fin; Cape Naturaliste northwards; Indo-E. Pacific; to 420 cm. All tackle world record 189.37 kg, Australian record 148 kg. ★★★

945 INDO-PACIFIC SAILFISH
Istiophorus platypterus (Shaw & Nodder)
Inhabits oceanic waters, generally well offshore; distinguished by sail-like dorsal fin; Cape Leeuwin northwards; Indo-E. Pacific; to 360 cm. All tackle world record 100.24 kg, Australian record 55.79 kg. ★★

946 SWORDFISH
Xiphias gladius Linnaeus
Inhabits oceanic waters, generally well offshore; distinguished by the long, flattened sword; all West Australian seas; worldwide in temperate and tropical seas; to 450 cm. All tackle world record 536.15 kg, Australian record 95.25 kg. ★★★

947 ALBACORE
Thunnus alalunga (Bonnaterre)
Inhabits oceanic waters, generally well offshore; occurs in schools; distinguished by the very long pectoral fins, these are shorter in juveniles and subadults which resemble 948 and 949, but differ from them by having a white rear border on the tail; all West Australian seas; worldwide in temperate and tropical seas; to 150 cm. All tackle world record 40 kg, Australian record 15 kg. ★★★

948 YELLOWFIN TUNA
Thunnus albacares (Bonnaterre)
Inhabits oceanic waters, generally well offshore; occurs in schools; distinguished by yellow dorsal and anal fins that become elongated with increased age; all West Australian seas; worldwide tropical and temperate seas; to 210 cm. All tackle world record 176.4 kg, Australian record 97 kg. ★★★

949 BIGEYE TUNA
Thunnus obesus (Lowe)
Inhabits oceanic waters, generally well offshore; occurs in schools; similar to 948, but has much larger eye and lacks elongate dorsal and anal rays; all West Australian seas; worldwide in tropical and temperate seas; to 240 cm. All tackle world angling record 197.3 kg, Australian record 37.13 kg. ★★

950 SOUTHERN BLUEFIN TUNA
Thunnus maccoyii (Castelnau)
Inhabits oceanic waters, generally well offshore; occurs in schools; distinguished by short pectoral fins and robust body shape; all West Australian seas; Southern Hemisphere in mainly temperate seas, but ranges to tropics in Australia; to 240 cm. All tackle world angling record 158 kg, Australian record 71.2 kg. ★★

951 NORTHERN BLUEFIN TUNA
Thunnus tonggol (Bleeker)
Inhabits oceanic waters, generally well offshore, occurs in schools; distinguished by short pectoral fins and slender body shape; Geographe Bay northwards; mainly W. Pacific and N. Indian Ocean; to 150 cm. All tackle world record 35.9 kg, Australian record 31.26 kg. ★★

952 SKIPJACK TUNA
Katsuwonis pelamis (Linnaeus)
Inhabits oceanic waters, generally well offshore; occurs in schools; distinguished by dark stripes on sides; all West Australian seas; worldwide in tropical and temperate seas. All tackle world record 18.93 kg, Australian record 9.3 kg. ★★

TONNES OF TUNA

The fishes featured on Plates 62-63 occur worldwide, primarily in tropical and temperate seas. They occur both inshore and far out to sea. Tunas, mackerels, and bonitos (family Scombridae; 49 species worldwide) are well known for their fine eating qualities. Over the past 10 years world catches have generally fluctuated between about 5 and 6 million tonnes annually. These fishes are also highly prized by recreational anglers and throughout much of the world they support important subsistence fisheries. All species are powerful swimmers and some undergo extensive annual migrations. The largest species is about 300 cm, but most grow to between 100-200 cm TL.

Billfishes exhibit similar habits and are close relatives of tunas, being distinguished by their elongate, spear-like snout. The group is comprised of the swordfish (Xiphidae; 1 species worldwide) and the marlins and sailfishes (Istiophoridae; 11 species worldwide). They are favourite angling fishes and some may reach the massive size of over 450 cm TL and 700 kg in weight.

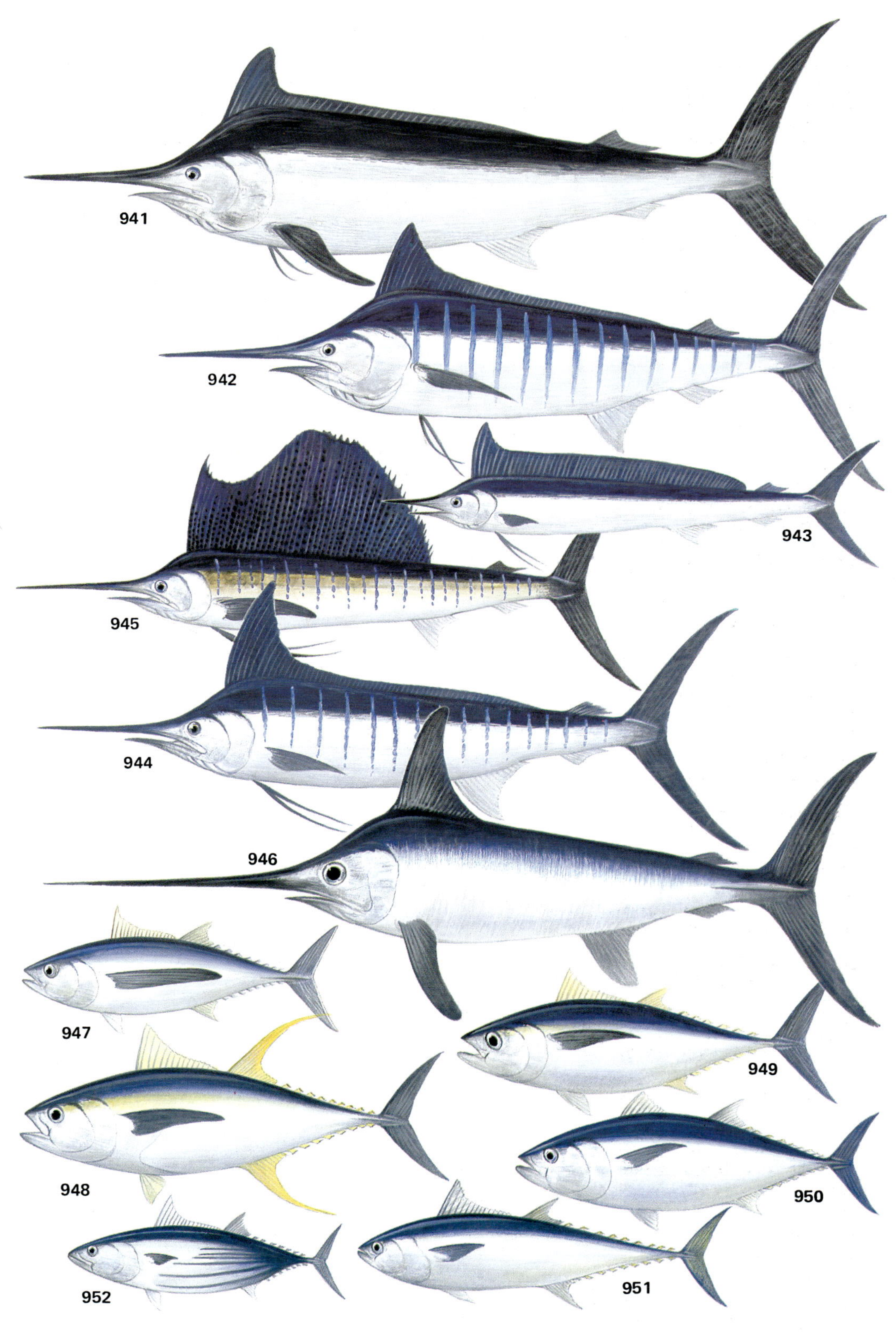

941
942
943
944
945
946
947
948
949
950
951
952

953 WAHOO
Acanthocybium solandri Cuvier
Inhabits oceanic waters, generally well offshore; occurs solitarily or in loose aggregations; distinguished from mackerels by its more elongate shape, more numerous spines in the first dorsal fin and by its banded pattern (except also in 954); North West Cape northwards; all tropical seas; to 210 cm. All tackle world record 67.6 kg Australian record 46.15 kg. ★★★

954 NARROW-BARRED SPANISH MACKEREL
Scomberomorus commerson (Lacepede)
Inhabits coastal seas, frequently near reefs; similar to 953, but has fewer dorsal spines (15-18 versus 23-27) and shorter first dorsal fin; Geographe Bay northwards; Indo-W. Pacific; to 235 cm. All tackle world record 44.9 kg Australian record 42.2 kg. ★★★★

955 BROAD-BARRED SPANISH MACKEREL
Scomberomorus semifasciatus (Macleay)
Inhabits coastal seas in the vicinity of reefs; distinguished by dark bars extending part way down sides and black area at front of dorsal fin; Shark Bay northwards; Australia-New Guinea; to 120 cm; 8.439 kg. ★★★

956 AUSTRALIAN SPOTTED MACKEREL
Scomberomorus munroi Collette & Russo
Inhabits coastal seas; distinguished by broad band of small spots along middle of sides; Abrolhos northwards; N. Australia only; to 104 cm; 10.206 kg. ★★★

957 QUEENSLAND SCHOOL MACKEREL
Scomberomorus queenslandicus Munro
Inhabits inshore coastal waters; distinguished by large dark spots on sides and black area at front of dorsal fin; Shark Bay northwards; N. Australia; to 100 cm; 12.247 kg. ★★★

958 ORIENTAL BONITO
Sarda orientalis (Temminck & Schlegel)
Inhabits coastal seas, sometimes in large schools; distinguished by narrow stripes on upper half of body; Great Australian Bight to Shark Bay; Indo-E. Pacific; to 102 cm; 3.5 kg. ★★

959 MACKEREL TUNA
Euthynnus affinis (Cantor)
Inhabits oceanic waters, sometimes well offshore; colour pattern is similar to 962, but distinguished by the lack of space between the dorsal fins; Cape Leeuwin northwards; Indo-C. Pacific; to 100 cm. All tackle world and Australian record 12.0 kg. ★★

960 LEAPING BONITO
Cybiosarda elegans (Whitley)
Inhabits coastal waters, sometimes entering estuaries; occurs in large schools; distinguished by spots on back and stripes on lower side, also tall, dark-coloured first dorsal fin; Geographe Bay northwards; Australia only (except south coast); to 54 cm; 1.15 kg. ★★

961 CORSELETTED FRIGATE MACKEREL
Auxis rochei Risso
Inhabits coastal and oceanic waters, forming large schools; distinguished by widely separated dorsal fins, elongate patch of short bars or blotches on back at rear half of body, and by its slender shape; all West Australian seas; worldwide tropical and subtropical seas; to 50 cm.

962 FRIGATE MACKEREL
Auxis thazard (Lacepede)
Inhabits coastal and oceanic waters, forming large schools; similar to 959, but has wide gap between dorsal fins and more slender shape; all West Australian seas; worldwide tropical and subtropical seas; to 58 cm; 4.536 kg. ★★

963 LONG-JAWED MACKEREL
Rastrelliger kanagurta (Cuvier)
Inhabits coastal waters, near reefs; forms large schools; distinguished by widely separate dorsal fins; narrow lines or rows of spots on upper part of body, and black spot near lower margin of pectoral fin, North West Cape northwards; Indo-W. Pacific; to 35 cm. ★★★

964 DOGTOOTH TUNA
Gymnosarda unicolor (Ruppell)
Inhabits offshore waters, usually in the vicinity of coral reefs; distinguished by large conical teeth, relatively large eye, and undulating lateral line; North West Shelf; Indo-C. Pacific; to 150 cm. All tackle world record 131 kg.

965 SHARK MACKEREL
Grammatorcynus bicarinatus (Quoy & Gaimard)
Inhabits offshore waters, usually in the vicinity of coral reefs; has double lateral line, similar to 966, but eye smaller and frequently has dark spots along belly; Geographe Bay northwards; Australia only (except south coast); to 130 cm; 11.566 kg. ★★

966 DOUBLE-LINED MACKEREL
Grammatorcynus bilineatus (Ruppell)
Inhabits offshore waters, usually in the vicinity of coral reefs; has double lateral line, similar to 965 but eye larger and lacks spotting along belly; Dampier Archipelago northwards; Indo-W. Pacific; to 70 cm. ★★

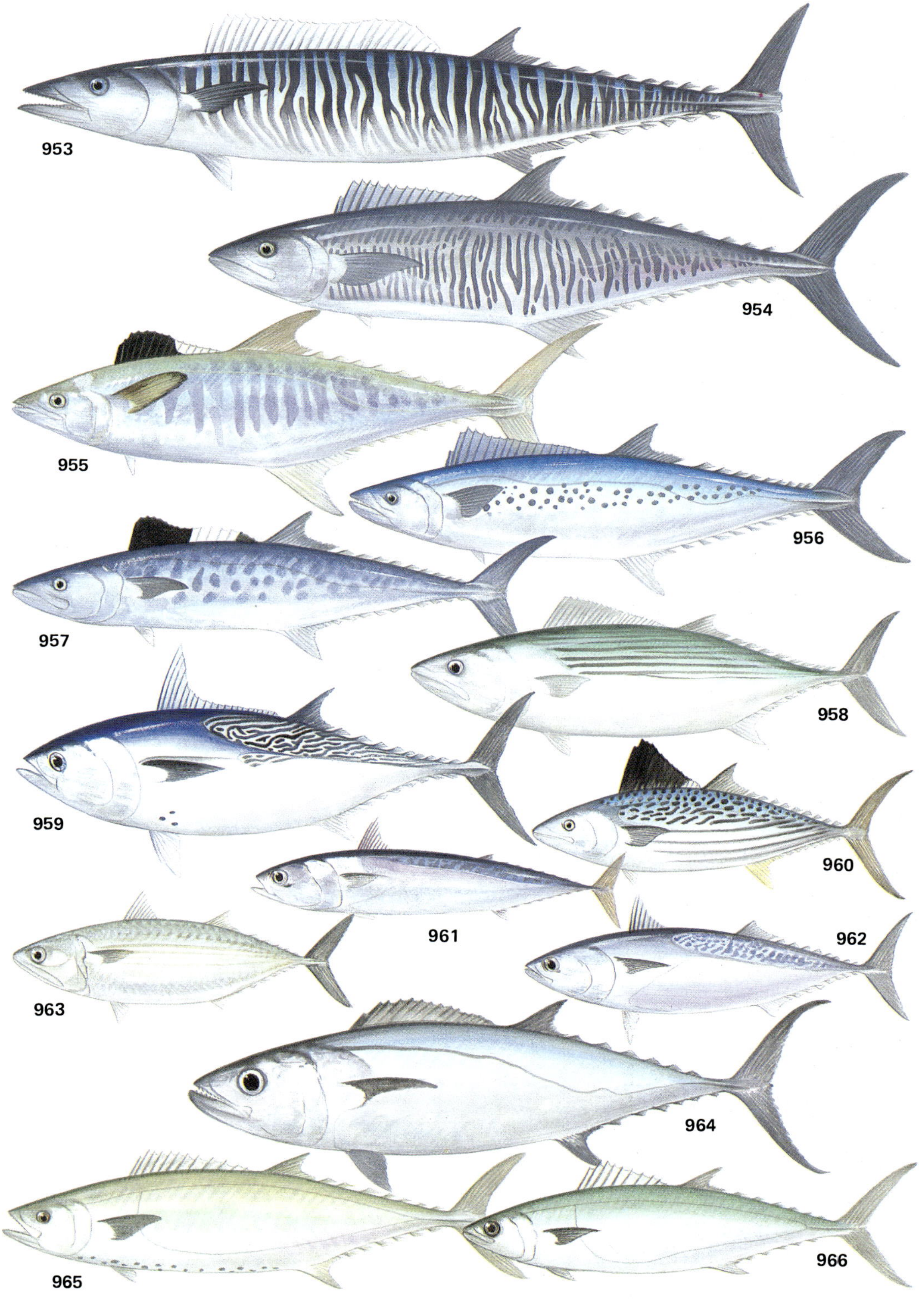

953

954

955

956

957

958

959

960

961

962

963

964

965

966

967 QUEENSLAND HALIBUT
Psettodes erumei (Bloch & Schneider)
Inhabits sand or mud bottoms; distinguished by large mouth and large sharp teeth; Mandurah northwards; Indo-W. Pacific; to 64 cm. ★★★

968 LARGE-TOOTHED FLOUNDER
Pseudorhombus arsius (Hamilton)
Inhabits sand or mud bottoms, sometimes in estuaries; distinguished by several enlarged teeth at front of mouth, large brown blotch behind pectoral fin, smaller blotch about half way between first spot and tail base, and scattered dark-edged pale spots; Fremantle northwards; Indo-W. Pacific; to 31 cm; 1.1 kg. ★★★

969 TWINSPOT FLOUNDER
Pseudorhombus diplospilus Norman
Inhabits sand bottoms; distinguished by 4 pairs of 'eye-spots'; other similar species (not shown) include *P. argus* (same 'eye-spot' pattern except 1 or 2 additional sets of spots on middle of body near tail base), *P. dupliocellatus* (same 'eye-spot' pattern as *P. diplospilus* but lacks enlarged canine teeth [versus strong canines]), *P. quinquocellatus* (similar pattern to *P. argus*, but only a single eye-spot instead of double spots in each marking) and *P. spinosus* (3 well separated, single spots on back); Onslow northwards; Indo-Australian Archipelago; to 25 cm. ★★

970 DEEP-BODIED FLOUNDER
Pseudorhombus elevatus Ogilby
Inhabits sand or mud bottoms; similar to 968, but lacks enlarged teeth, is deeper-bodied (i.e. rounder), and generally has more pale spots; Exmouth Gulf northwards; N. Indian Ocean and Indo-Australian Archipelago; to 15 cm.

971 SMALL TOOTHED FLOUNDER
Pseudorhombus jenynsii (Bleeker)
Inhabits sand or mud bottoms; distinguished by 6 pale-edged spots arranged : : . . ; entire coast of W.A.; Australia only; to 34 cm; 1.471 kg. ★★★

972 INTERMEDIATE FLOUNDER
Asterhombus intermedius (Bleeker)
Inhabits sand bottoms; distinguished by dense scattering of dark speckles and blotches on fins and body; also known as Blotched flounder; Exmouth Gulf northwards; Indian Ocean and Indo-Australian Archipelago; to 13 cm.

973 SPINY-HEADED FLOUNDER
Engyprosopon grandisquama (Temminck & Schlegel)
Inhabits sand bottoms; distinguished by well separated eyes and pair of black spots on margin of tail; also known as Mottled wide-eyed flounder; Abrolhos northwards; Indo-W. Pacific; to 13 cm.

974 THREE-SPOT FLOUNDER
Grammatobothus polyophthalmus (Bleeker)
Inhabits sand bottoms; distinguished by mottled pattern and 3 pale-edged dark spots; Exmouth Gulf northwards; N.E. Indian Ocean and Indo-Australian Archipelago; to 14 cm.

975 PANTHER FLOUNDER
Bothus pantherinus (Ruppell)
Inhabits sand bottoms near coral reefs; distinguished by mottled appearance with numerous brown and white spots, and large brown patch in middle of rear part of body, males have filamentous pectoral rays reaching tail base; also known as Leopard flounder; Abrolhos northwards; Indo-W. Pacific; to 24 cm. ★★

976 DARK THICK-RAYED SOLE
Aesopia cornuta Kaup
Inhabits sand or mud bottoms; similar to 980 and 981, but has darker stripes; *A. heterorhinos* (not shown) also similar but has irregular-shaped broken bars and white black-tipped tail; Dampier northwards; N. Australia only; to 18 cm.

977 TUFTED SOLE
Dexillichthys muelleri (Steindachner)
Inhabits sand bottoms; distinguished by light brown colour with patches of dark skin filaments; Rottnest Island northwards; Australia only; to 20 cm.

978 DARK-SPOTTED SOLE
Aseraggodes melanospilus (Bleeker)
Inhabits sand bottoms; distinguished by numerous small dark spots and pair of large dark blotches superimposed on mottled pattern. Shark Bay northwards; mainly W. Pacific, to 15 cm.

979 PEACOCK SOLE
Pardachirus pavoninus (Lacepède)
Inhabits sand bottoms; distinguished by numerous black spots with broad pale margins; experiments have shown that the mucus of this fish has shark-repellant qualities; Onslow northwards; Indo-C. Pacific; to 25 cm.

980 WICKER-WORK SOLE
Zebrias craticula (McCulloch)
Inhabits sand bottoms; similar to 976 and 981, but dark bars more numerous and narrower; Shark Bay northwards; N. Australia only; to 15 cm.

981 HARROWED SOLE
Strabozebrias cancellatus (McCulloch)
Inhabits sand bottoms; bars lighter and body more elongate than 976; bars wider and less numerous than 980; Albany northwards; Australia only; to 27 cm; .14 kg.

982 SPOTTED SOLE
Phyllichthys punctatus McCulloch
Inhabits sand bottoms; distinguished by dense covering of diffuse white blotches; Fremantle northwards; Australia only; to 24 cm.

983 COCKATOO FLOUNDER
Samaris cristatus Gray
Inahbits sand bottoms; distinguished by filamentous dorsal fin-rays on top of head; also known as Cockatoo righteye flounder; Port Hedland northwards; Indo-W. Pacific; to 17 cm.

984 OCELLATED FLOUNDER
Psammodiscus ocellatus Günther
Inhabits sand bottoms; distinguished by dense mottled or speckled pattern 3 or 4 paled edged dark spots usually present; Exmouth Gulf northwards; Indo-Australian Archipelago; to 15 cm.

985 PATTERNED TONGUE SOLE
Paraplagusia bilineata (Bloch)
Inhabits sand bottoms; distinguished by elongate shape and fringe of branched tentacles on lips; other elongate soles (not shown) from this area without fringe of tentacles belong to the genus *Cynoglossus*; Geraldton northwards; Indo-W. Pacific; to 40 cm. ★★★

986 SPOTTED DUCKBILL
Bembrops aethalea McKay
Inhabits offshore trawling grounds; distinguished by 2 spines on gill cover, a loose flap of skin on rear part of upper jaw, and small black spots on back; Broome northwards; N.W. Australia only; to 20 cm.

987 SHARPNOSED DUCKBILL
Bembrops filodorsalia Okada & Suzuki
Inhabits offshore trawling grounds; similar to 986, but lacks distinct spotting and tail has dark margin with dark spot on upper part of base; Broome northwards; mainly W. Pacific; to 20 cm.

988 BLOTCHED DUCKBILL
Chironema chlorotaenia McKay
Inhabits offshore trawling grounds; distinguished by 2 spines on gill cover, no flap of skin at rear of upper jaw, elongate orange spots, and diffuse brown blotches on sides; Broome northwards; N.W. Australia only; to 21 cm.

989 INDIAN DRIFTFISH
Ariomma indica (Day)
Inhabits oceanic waters usually well offshore; distinguished by ovate silvery body and rounded snout with small mouth; similar to 990, but has 2 dorsal fins; Onslow northwards; Indo-W. Pacific; to 25 cm.

990 NORTH-WEST RUFFE
Psenopsis humerosa Munro
Inhabits oceanic waters usually well offshore; similar to 989, but has only one dorsal fin; Dampier northwards; N.W. Australia only; to 20 cm.

991 WHITELEGGE'S EYEBROWFISH
Cubiceps whiteleggii (Waite)
Inhabits oceanic waters usually well offshore; distinguished by elongate shape, 2 dorsal fins and overall darkish colour; Broome northwards; Australia only; to 13 cm.

992 THREE-TOOTH PUFFER
Triodon macropterus Lesson
Inhabits coastal waters and offshore trawling grounds; distinguished by large skin flap on belly and black spot on middle of side; Onslow northwards; Indo-W. Pacific; to 60 cm. Ⓟ

993 BLACKTAIL TRIPODFISH
Triacanthus biaculeatus (Bloch)
Inhabits trawling grounds; similar to 994, but spiny dorsal fin entirely black (versus only front half of fin black); Exmouth Gulf northwards; Indo-W. Pacific; to 25 cm.

994 SILVER TRIPODFISH
Triacanthus nieuhofi Bleeker
Inhabits trawling grounds; similar to 993, but rear half of spiny dorsal fin pale (versus black); Broome northwards; mainly Indo-Australian Archipelago; to 28 cm.

995 BLACKTIP TRIPODFISH
Trixiphichthys weberi (Chadhuri)
Inhabits trawling grounds; distinguished by black tip on spiny dorsal fin and diffuse dark blotches on side, similar to 996, but has longer, narrower snout; Onslow northwards; Andaman Sea and Indo-Australian Archipelago; to 20 cm.

996 BLOTCHED TRIPODFISH
Pseudotriacanthus strigilifer (Cantor)
Inhabits trawling grounds; similar to 995, but has more pronounced pattern of blotches on side and shorter snout; Dampier northwards; N. Indian Ocean and Indo-Australian Archipelago; to 24 cm.

997 LONGSNOUT SPIKEFISH
Halimochirurgus centriscoides Alcock
Inhabits trawling grounds; similar to 999, but snout is directed upwards and lacks enlarged mouth opening at tip; Broome northwards; Andaman Sea and Indo-Australian Archipelago; to 15 cm.

998 SHORTSNOUT SPIKEFISH
Triacanthodes ethiops Alcock
Inhabits trawling grounds; similar in general shape to 993-996, but has 6 spines (versus 5) in first dorsal fin and much shorter and thicker tail base, usually has 3 diverging stripes on side; Broome northwards; Indo-W. Pacific; to 9 cm.;

999 TRUMPETSNOUT SPIKEFISH;
Macrorhamphosodes platycheilus Fowler
Inhabits trawling grounds; similar to 997, but has snout directed straight forward and enlarged mouth opening at tip; Broome northwards; Indian Ocean and Indo-Australian Archipelago; to 13 cm.

TRASH FISHES

Most of the species featured on this plate, particularly the duckbills, tripodfishes and spikefishes, are classed as 'trash' by commercial fishermen. These species, along with many other small bottom fishes are often captured in large numbers while trawling. Because of their small size or poor edibility they are dumped overboard after sorting out the marketable species. Unfortunately most of the trash catch dies during this process, resulting in tremendous waste of this resource. However, in some parts of the world the trash catch is effectively harvested for use as fish meal and fertilizer.

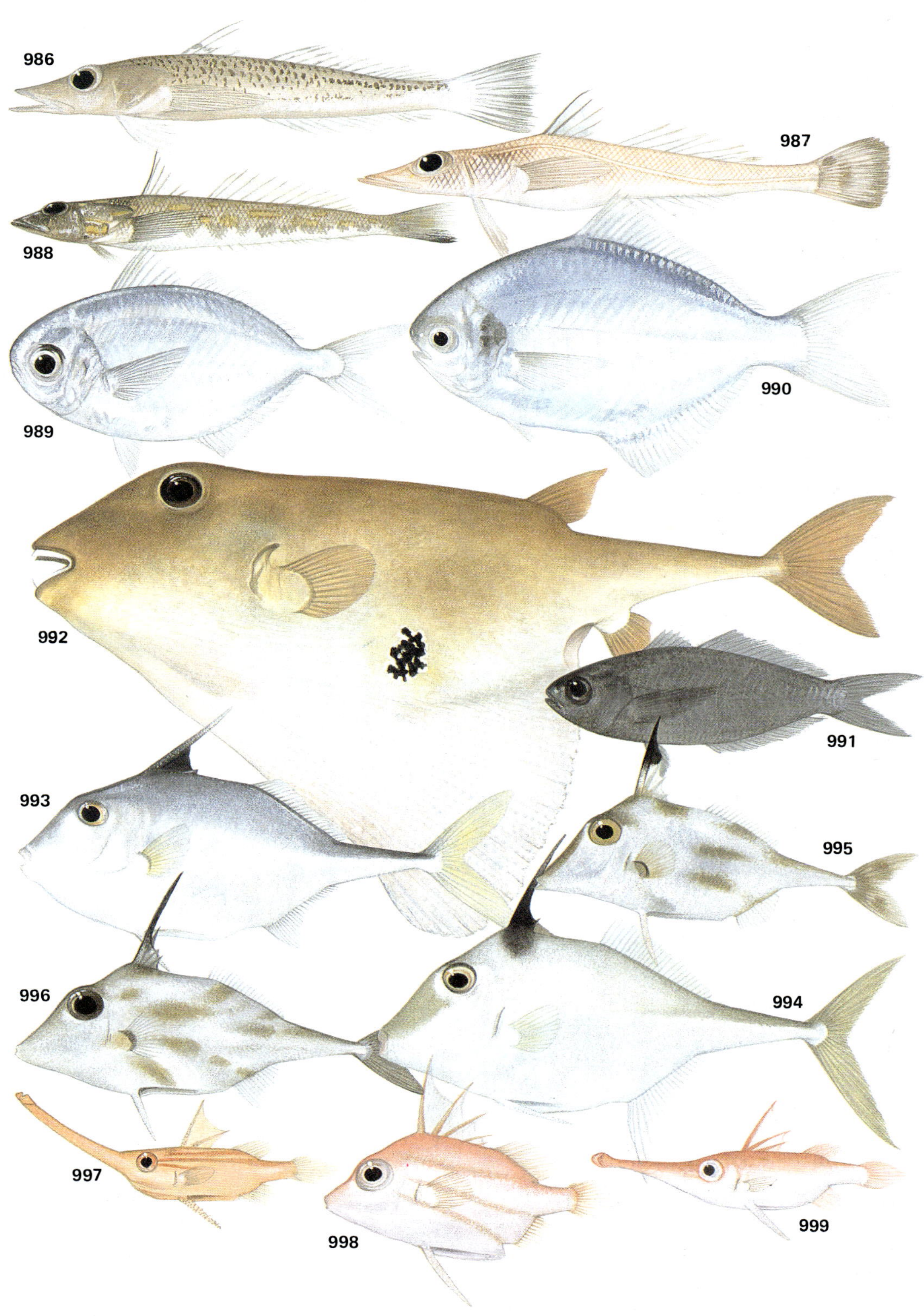
986
987
988
989
990
991
992
993
994
995
996
997
998
999

1000 RED-LINED TRIGGERFISH
Balistapus undulatus (Park)
Inhabits coral reefs; distinguished by red diagonal stripes; Ningaloo Reef northwards; Indo-C. Pacific; to 35 cm.

1001 STARRY TRIGGERFISH
Abalistes stellatus (Bloch & Schneider)
Inhabits coral and rocky reefs; distinguished by relatively elongate shape, blue to orange spotting on body, and frequently with 3 white patches on back at base of dorsal fin; large adults have elongate filaments on tail; Shark Bay northwards; Indo-W. Pacific; to 60 cm; 2.4 kg.

1002 BLUE-FINNED TRIGGERFISH
Balistoides viridescens (Bloch & Schneider)
Inhabits coral and rocky reefs; nest-guarding males may attack divers inflicting painful bites; distinguished by relatively large size, dark area with pale 'bridle' around upper part of mouth, yellowish cheeks, diffuse dark bar through eye and dark fin margins; Rottnest Island northwards; Indo-C. Pacific; to 60 cm; 2.3 kg.

1003 EBONY TRIGGERFISH
Melichthys niger (Bloch)
Inhabits coral reefs; distinguished by black colour of body and fins; *Melichthys vidua* (p. 158) is a similar species, but has pale dorsal, anal, and tail fins; Ningaloo Reef northwards; Indo-C. Pacific; to 35 cm.

1003a PINKTAIL TRIGGERFISH
Melichthys vidua (Solander)
See Plate 70.

1004 WHITE-BARRED TRIGGERFISH
Rhinecanthus aculeatus (Linnaeus)
Inhabits coral reefs; distinguished by white diagonal bars above anal fin base; Point Quobba northwards; Indo-C. Pacific; to 30 cm; .44 kg.

1005 WEDGE-TAILED TRIGGERFISH
Rhinecanthus rectangulus (Day)
Inhabits coral reefs on limestone platforms exposed to wave action; distinguished by black wedge-shaped mark at tail base; Ningaloo Reef northwards; Indo-C. Pacific; to 24 cm.

1006 YELLOW-SPOTTED TRIGGERFISH
Pseudobalistes fuscus (Bloch & Schneider)
Inhabits coral reefs; distinguished by dense network of yellow-orange spots on body and pointed lobes of tail; Ningaloo Reef northwards; Indo-C. Pacific; to 50 cm.

1007 BLACK TRIGGERFISH
Sufflamen chrysopterus (Bloch & Schneider)
Inhabits coral reefs; distinguished by dark colour of body, pale line below eye and white-edged tail; Ningaloo Reef northwards; Indo-C. Pacific; to 30 cm; .312 kg.

1008 PALLID TRIGGERFISH
Sufflamen bursa (Bloch & Schneider)
Inhabits coral reefs, often in rubble areas, distinguished by general grey colouration with narrow brown to orange bar through eye and another from dorsal fin to pectoral base; Coral Bay northwards; Indo-C. Pacific; to 30 cm.

1009 BROWN TRIGGERFISH
Sufflamen frenatus (Latreille)
Inhabits coral reefs and hard flat-bottom areas with occasional outcrops; distinguished by brown colour and bridle-like marking behind mouth; Shark Bay northwards; Indo-C. Pacific; to 30 cm; .835 kg.

1010 LINED TRIGGERFISH
Xanthichthys lineopunctatus (Hollard)
Inhabits offshore reefs, often seen well above the bottom; distinguished by narrow lines on head and body and orange outline on tail; North West Shelf; Indo-W. Pacific; to 30 cm.

TRIGGERFISHES

Triggerfishes (family Balistidae) are characterised by a football-like shape, leathery skin, and small mouth with powerful crushing jaws. They are mainly inhabitants of the Indo-Pacific region, but also occur in other warm seas. There are an estimated 35 species worldwide. Some of the larger triggers, particularly 1002 are a menace to divers at certain times of the year when guarding nests. The male parent is particularly vicious and may aggressively charge other large fishes or humans. Although the mouth is small they are capable of delivering a painful bite. At night triggerfishes tightly wedge themselves in coral crevices by extending the dorsal spines which are held erect by a peculiar locking mechanism. Some species, for example 1004 and 1005, are capable of producing grunt-like sounds when disturbed. They feed on a wide variety of intertebrates including sponges, gorgonians, hydroids, corals, crabs, shrimps, molluscs, and echinoderms.

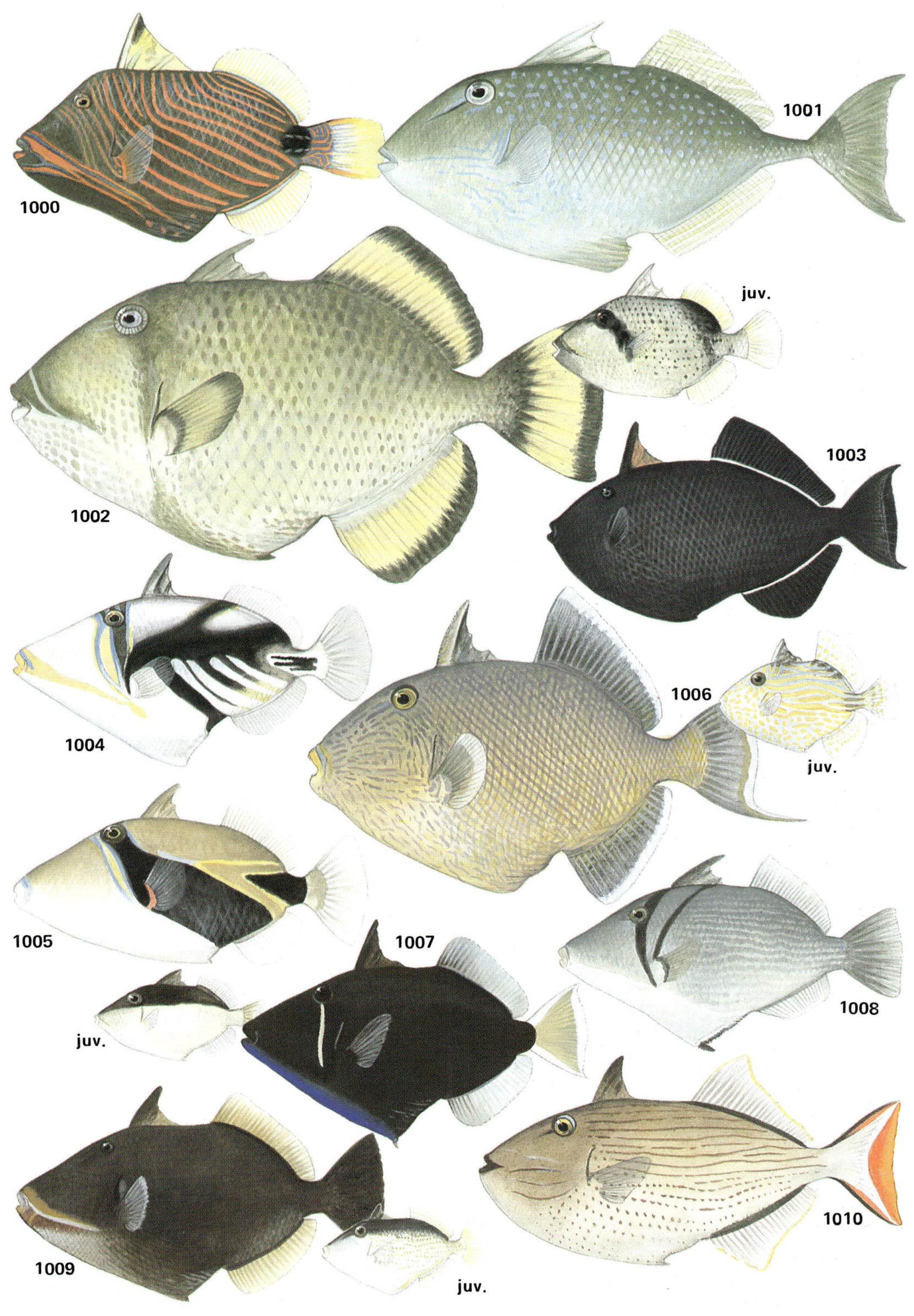

1000
1001
1002
juv.
1003
1004
1005
1006
juv.
1007
1008
juv.
1009
1010
juv.

1011 UNICORN LEATHERJACKET
Aluterus monoceros (Linnaeus)
Inhabits flat bottom trawling grounds; distinguished by large ovate shape and slender dorsal spine, similar to 1012, but has rounder head, shorter tail and lacks ornate pattern; Fremantle northwards; worldwide tropical and subtropical seas; to 76 cm; 1.95 kg. ★★★

1012 SCRIBBLED LEATHERJACKET
Aluterus scriptus (Osbeck)
Inhabits coral reefs; distinguished by relatively elongate shape, slender dorsal spine, spotted pattern, and elongate tail; Point Cloates northwards; worldwide tropical and subtropical seas; to 100 cm. ★★★

1013 HONEYCOMB LEATHERJACKET
Cantherhines pardalis (Rüppell)
Inhabits coral reefs; distinguished by network of dark spots on sides; Point Quobba northwards; Indo-C. Pacific; to 20 cm. ★★★

1014 SPECTACLED LEATHERJACKET
Cantherhines fronticinctus (Günther)
Inhabits coral reefs; distinguished by white bar on tail base, sometimes with irregular dark blotches on sides; Ningaloo Reef northwards; Indo-C. Pacific; to 23 cm. ★★★

1015 BEARDED LEATHERJACKET
Anacanthus barbatus Gray
Inhabits sand-weed bottoms; distinguished by unusually elongate shape and barbel on chin; Albany northwards, but rare south of Shark Bay; E. Indian Ocean and Indo-Australian Archipelago; to 35 cm.

1016 BLUE-SPOTTED LEATHERJACKET
Eubalichthys caeruleoguttatus Hutchins
Inhabits trawling grounds; male has distinctive pattern of stripes on head and body and elevated dorsal and anal fin lobes, female distinguished by blue to whitish spots on side; Abrolhos northwards; W.A. only; to 38 cm. ★★★

1017 FAN-BELLIED LEATHERJACKET
Monacanthus chinensis (Osbeck)
Inhabits reef and weed bottoms, also trawling grounds; distinguished by triangular back profile, huge skin flap on belly, and filamentous extension on tail; Geographe Bay northwards; mainly W. Pacific; to 38 cm; .58 kg. ★★★

1018 PRICKLY LEATHERJACKET
Chaetoderma penicilligera (Cuvier)
Inhabits sea grass and trawling grounds; distinguished by round shape and tentacles on head and body; Geographe Bay northwards; mainly W. Pacific; to 31 cm. ★★★

1019 BEAKED LEATHERJACKET
Oxymonacanthus longirostris (Bloch & Schneider)
Inhabits coral reefs, usually associated with branching or plate corals; distinguished by tubular snout and orange spots; Abrolhos northwards; Indo-W. Pacific; to 9 cm.

1020 THREADFIN LEATHERJACKET
Paramonacanthus filicauda (Günther)
Inhabits trawling grounds; distinguished by relatively deep body, dark blotch below front of dorsal fin, and filament on upper lobe of tail; Barrow Island northwards; Australia only; to 22 cm. ★★★

1021 HAIR-FINNED LEATHERJACKET
Paramonacanthus choirocephalus (Bleeker)
Inhabits trawling grounds; distinguished by slightly diagonal dark stripes on side with intense blotch in middle stripe just behind pectoral fin; female shown here, male is much more slender; Abrolhos northwards; N.W. Australia and Indo-Malayan region; to 14 cm. ★★★

1022 RED-TAILED LEATHERJACKET
Pervagor janthinosoma (Bleeker)
Inhabits coral reefs; distinguished by brownish body and fan-shaped red tail; Lancelin northwards; Indo-W. Pacific; to 14 cm.

1023 MODEST LEATHERJACKET
Thamnaconus modestoides (Barnard)
Inhabits trawling grounds; distinguished by relatively elongate shape and non-descript pattern; Kalbarri northwards; Indian Ocean and W. Pacific; to 30 cm.

1024 BROWN-BLOTCHED LEATHER-JACKET
Stephanolepis sp.
Inhabits trawling grounds; an undescribed species distinguished by a relatively deep body and dense pattern of dark blotches; Shark Bay to North West Cape, juveniles range to Rottnest Island; W.A. only; to 34 cm. ★★★

1025 POT-BELLIED LEATHERJACKET
Pseudomonacanthus peroni (Hollard)
Inhabits trawling grounds; distinguished by moderately large skin flap on belly and small spots on body and tail; Rottnest Island northwards; Australia only; to 40 cm. ★★★

1026 PAXMAN'S LEATHERJACKET
Colurodontis paxmani Hutchins
Inhabits sea grass; distinguished by movable pelvic spine and flattened teeth at front of jaws; Fremantle northwards; but rare south of Shark Bay; Australia only; to 15 cm.

LEATHER JACKETS

Leatherjackets (family Monacanthidae) are closely related to triggerfishes (Plate 66) and share many of their anatomical peculiarities. However they are generally more laterally compressed and usually have two dorsal spines instead of three. Members of both groups are able to swim slowly by undulating movements of the soft dorsal and anal fins. Rapid bursts are achieved mainly by vigorous tail movement. Australia has more leatherjackets than any other region with nearly 60 of the estimated total of 85 species being represented. However, the majority are confined to temperate and subtropical seas. The flesh of many species is good eating and in places such as Australia and Japan they are important commercial fishes. Food habits are similar to those of triggerfishes and consist mainly of benthic invertebrates.

1011
1013
1012
1014
1015
1017
1016
1019
1018
1022
1020
1021
1026
1023
1025
1024

1027 LONG-HORNED COWFISH
Lactoria cornuta (Linnaeus)
Inhabits weed-sand areas near rock or coral reefs; distinguished from 1028 by longer horns and flatter belly that is not semi-transparent; *Lactoria fornasini* (not shown) also has long horns, but has a well developed hump on back; Fremantle northwards; Indo-C. Pacific; to 46 cm. ℗

1028 ROUNDBELLY COWFISH
Lactoria diaphana (Bloch & Schneider)
Inhabits coastal waters, the young sometimes in estuaries; distinguished from 1027 by shorter horns; "thorn" on back, and semi-transparent, rounded belly; Shark Bay northwards; Indo-W. Pacific; to 25 cm. ℗

1029 YELLOW BOXFISH
Ostracion cubicus (Linnaeus)
Inhabits coral reefs and southern rocky reefs; distinguished by yellow to brown colour with small black spots arranged in clusters on side, juvenile bright yellow with black spots; Fremantle northwards; Indo-W. Pacific; to 45 cm. ℗

1030 SPOTTED BOXFISH
Ostracion meleagris (Linnaeus)
Inhabits coral reefs; male distinguished by blue sides with yellow-orange spots, female by black colour and numerous white spots; Ningaloo Reef northwards; Indo-C. Pacific; to 16 cm. ℗

1031 HORN-NOSED BOXFISH
Rhynchostracion rhinorhynchus (Bleeker)
Inhabits reefs and flat bottom areas; distinguished by bump on snout; Shark Bay northwards; mainly Indo-Australian Archipelago; to 35 cm. ℗

1032 SMALL-NOSED BOXFISH
Rhynchostracion nasus (Bloch)
Inhabits reefs and flat bottom areas; distinguished by lack of horns, a ridged back, and dull spots on side; Shark Bay northwards; Indo-W. Pacific; to 30 cm. ℗

1033 BLACK-BLOTCHED TURRETFISH
Tetrasomus gibbosus (Linnaeus)
Inhabits coastal waters, sometimes trawled; distinguished by prominent peak on back; Dampier Archipelago northwards; Indo-W. Pacific; to 30 cm. ℗

1034 TURRETFISH
Trioris reipublicae (Ogilby)
Apparently a deeper water fish that is sometimes washed ashore during storms; distinguished by ridge on back with 2 stout spines, and blue lines on body, young have peak on back similar to 1033, but with 2 spines at apex; Albany northwards; Australia and New Guinea; to 22 cm. ℗

1035 FINE-SPINED PUFFERFISH
Tylerius spinosissimus (Regan)
Inhabits trawling grounds; distinguished by short bristles covering body; North West Cape northwards; Indo-W. Pacific; to 12 cm. ℗

1036 MANY-STRIPED PUFFERFISH
Anchisomus multistriatus Richardson
Inhabits trawling grounds and offshore reefs; distinguished by curved bars on top half of head and body with spots below; North West Cape northwards; mainly S.W. Pacific; to 15 cm. ℗

1037 NARROW-LINED TOADFISH
Arothron manillensis de Procé
Inhabits relatively turbid inshore waters over silty sand or mud bottoms, sometimes amongst weeds; distinguished by numerous thin stripes on side; Exmouth Gulf northwards; mainly W. Pacific; to 31 cm. ℗

1038 RETICULATED PUFFERFISH
Arothron reticularis (Bloch & Schneider)
Inhabits shallow coastal waters, usually over sand or mud bottoms; similar to 1039, but with dark lines on snout and cheek; Dampier Archipelago northwards; W. Pacific and E. Indian Ocean; to 30 cm. ℗

1039 STARS AND STRIPES TOADFISH
Arothron hispidus (Linnaeus)
Inhabits shallow waters near rock or coral reefs; similar to 1038, but without lines on snout and cheek; Fremantle northwards; Indo-C. Pacific; to 51 cm; 3.5 kg. ℗

1040 IMMACULATE PUFFERFISH
Arothron immaculatus (Bloch & Schneider)
Inhabits sand or mud bottoms, often amongst weed; distinguished by lack of markings and dark rim around tail; not yet recorded from the northwest but likely to occur there; Indo-C. Pacific; to 30 cm. ℗

1041 STARRY PUFFERFISH
Arothron stellatus (Bloch & Schneider)
Inhabits sand and mud bottoms, often amongst weeds; distinguished by large size and dense spotting on head and body, juvenile with curved black bars on belly; Shark Bay northwards; Indo-C. Pacific; to 90 cm. ℗

1042 BLACK-SPOTTED TOADFISH
Arothron nigropunctatus (Bloch & Schneider)
Inhabits clear waters of offshore coral reefs; distinguished by scattered black spots and dark lips, background colour ranges from grey to yellow; North West Shelf; Indo-C. Pacific; to 30 cm. ℗

BOXFISHES AND BLOWIES

Boxfishes of the family Ostraciidae (1027-1034) are strange creatures found mainly on tropical and subtropical reefs. Their bodies are encased in a bony carapace and the fins are relatively small. They usually are slow swimmers, but are capable of short, rapid bursts. When feeding boxfishes sometimes squirt a jet of water into the sand to uncover small plants and invertebrate animals that are then sucked into the mouth. Some boxfishes produce a toxic mucus that can kill other fishes or even themselves when confined to a small aquarium.

The puffers or blowies (family Tetraodontidae) and the related porcupinefishes (family Diodontidae) can inflate their body by swallowing water (or air if out of water). Presumably this adaptation serves as a deterrent to potential predators. The estimated 140 species of puffers occur worldwide in tropical and warm temperate seas and estuaries. The flesh, especially the viscera, contains a potent toxin that has caused many human fatalities. However, they are considered to be a great delicacy in Japan, where they are prepared by specially trained and licensed cooks. The largest species grow to about 1 m TL, but most are considerably less.

1027
1029
1028
1030
♂
♀
1031
1032
1035
1033
1034
1036
1037
1038
1039
1040
1041
juv.
1042

1043 THREE-BARRED PUFFER
Canthigaster coronata (Vaillant & Sauvage)
Inhabits flat bottom areas with coral, sponge, and rocky outcrops; distinguished by 3 dark saddles on upper half of body; Exmouth Gulf northwards; Indo-C. Pacific; to 13 cm. ℗

1044 SPOTTED PUFFER
Canthigaster janthinoptera (Bleeker)
Inhabits caves and crevices of coral reefs; distinguished by dense network of white lines and spots on head and body; Ningaloo Reef northwards; Indo-C. Pacific; to 8.5 cm. ℗

1045 BROWN-LINED PUFFER
Canthigaster rivulata (Schlegel)
Inhabits offshore reefs and trawling grounds; distinguished by wavy lines on back and brown stripe on side that curves around front of pectoral fin base; Shark Bay northwards; Indo-W. Pacific; to 20 cm. ℗

1046 MILK-SPOTTED TOADFISH
Chelonodon patoca (Hamilton)
Inhabits bays and brackish mangrove estuaries; distinguished by network of large dark-centred white spots on head and body; Onslow northwards; Indo-W. Pacific; to 20 cm. ℗

1047 ROUGH GOLDEN TOADFISH
Lagocephalus lunaris (Bloch & Schneider)
Inhabits coastal waters; similar to 1048 and 1049, but has bristles on dorsal surface between snout and dorsal fin; North West Cape northwards; Indo-W. Pacific; to 30 cm. ℗

1048 SMOOTH GOLDEN TOADFISH
Lagocephalus inermis (Temminck & Schlegel)
Inhabits coastal waters; similar to 1047 and 1049, but lacks bristles on dorsal surface (and elsewhere); North West Cape northwards; Indo-W. Pacific; to 20 cm. ℗

1049 BROWN-BACKED TOADFISH
Lagocephalus spadiceus (Richardson)
Inhabits coastal waters; similar to 1047 and 1048, but has patch of bristles on dorsal surface from snout to about half way to dorsal fin; North West Cape; Indo-W. Pacific; to 30 cm. ℗

1050 SILVER TOADFISH
Lagocephalus scleratus (Gmelin)
Inhabits coastal waters, usually in schools; an aggressive species that can inflict painful bites, in a feeding frenzy it is very dangerous, attacking everything in sight; distinguished by spots and faint blotches on back and silvery stripe on sides, is also more elongate than 1047-1049; Geographe Bay northwards; Indo-W. Pacific; to 85 cm; 6.5 kg. ℗

1051 DARWIN'S TOADFISH
Marilyna darwinii (Castelnau)
Inhabits mud bottom areas, frequently in mangrove estuaries or the lower reaches of freshwater streams; Derby northwards; N. Australia and S. New Guinea; to 17 cm. ℗

1052 HICK'S TOADFISH
Torquigener hicksi Hardy
Inhabits coastal waters; similar to 1053 and 1054, but bristles sparsely distributed on head and body; North West Cape northwards; N. Australia only; to 13 cm. ℗

1053 YELLOW-EYED TOADFISH
Torquigener parcuspinus Hardy
Inhabits coastal waters; similar to 1052, but covered with well developed bristles, lacks distinctive spotting of 1054 which has shorter bristles, also distinguished by small yellow areas above each eye; North West Cape northwards; N. Australia only; to 10 cm. ℗

1054 ORANGE-SPOTTED TOADFISH
Torquigener pallimaculatus Hardy
Inhabits coastal waters; distinguished by pattern of spots and bristle development that is intermediate to 1052 and 1053, also by orange-brown spots on lower side; Rottnest Island northwards, rare south of Shark Bay; N. Australia only; to 15 cm. ℗

1055 FRECKLED PORCUPINEFISH
Diodon holacanthus (Linnaeus)
Inhabits vicinity of coral reefs and southern rocky reefs; distinguished by long movable spines on head and body, and combination of scattered small dark spots and larger dark blotches (often much darker than shown); *Diodon hystrix* (not shown) is similar but lacks the larger dark patches, and the small spots are more numerous; Fremantle northwards; worldwide in tropical seas; to 35 cm. ℗

1056 BLOTCHED PORCUPINEFISH
Diodon liturosus Shaw
Inhabits vicinity of coral reefs; distinguished by long movable spines on head and body, and large dark patches on side and dorsal surface, a solid broad bar across the top of the head just behind eyes is also distinctive; Ningaloo Reef northwards; mainly W. and C. Pacific; to 40 cm. ℗

1057 SPOTBASE BURRFISH
Chilomycterus spilostylus Leis & Randall
Inhabits coastal waters in the vicinity of reefs; distinguished by short non-movable spines on head and body, and small back spots at the base of most spines; North West Cape northwards; N. Indian Ocean and S. China Sea; to 35 cm. ℗

1058 SPOTFIN PORCUPINEFISH
Chilomyterus reticulatus (Linnaeus)
Inhabits coastal waters in the vicinity of reefs; distinguished by short non-movable spines on head and body, faint dark bars on side, and spotted tail; Broome northwards; worldwide in tropical seas; to 55 cm. ℗

1059 LONG-SPINED PORCUPINEFISH
Cyclichthys jaculiferus (Cuvier)
Inhabits coastal waters; distinguished by long non-movable spines, 3-4 dark patches on back, and a few dark spots on side; Derby northwards; N. Australia and Arafura Sea; to 30 cm. ℗

1060 SHORT-SPINED PORCUPINEFISH
Cyclichthys orbicularis (Bloch)
Inhabits coastal waters; distinguished by short non-movable spines on head and body, and relatively large black spots on back and side; Shark Bay northwards; Indo-W. Pacific; to 30 cm. ℗

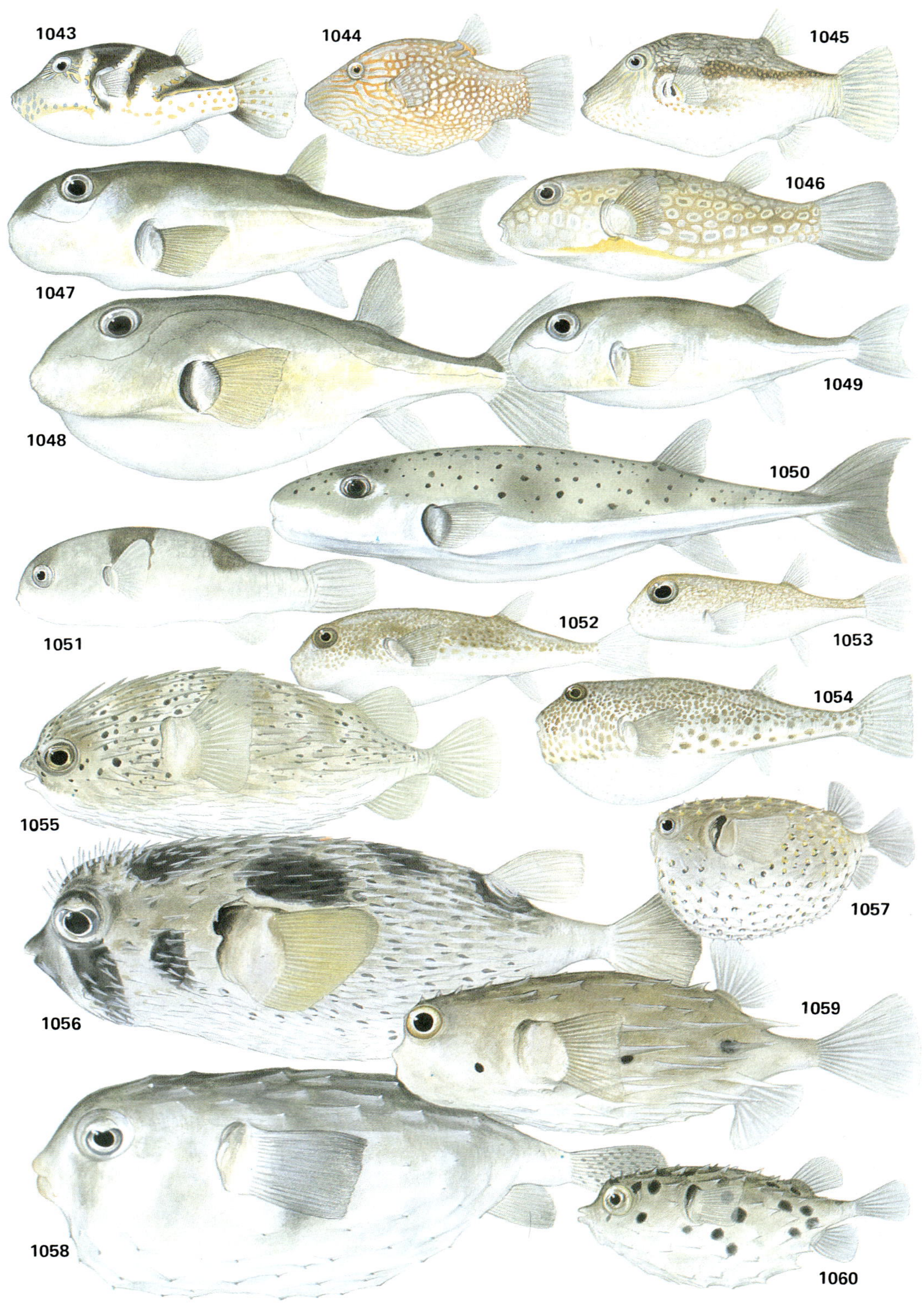
1043
1044
1045
1046
1047
1048
1049
1050
1051
1052
1053
1054
1055
1056
1057
1058
1059
1060

NOTE — After the paintings and text for the book were completed several important omissions were discovered. These have been included on this plate.

509 GOLD-SPOTTED SWEETLIPS
Plectorhinchus flavomaculatus (Ehrenberg)
Inhabits coral reefs and southern rocky reefs; similar to 510, but lacks yellow on fins and has small dark spots on dorsal fin and tail; Geographe Bay northwards; Indo-W. Pacific; to 50 cm; 2.96 kg. ★★

530 YELLOW-TAILED EMPEROR
Lethrinus atkinsoni Seale
Inhabits coral reefs; distinguished by steep forehead and yellow area along middle of sides extending on to tail; formerly confused with *L. mahsena* from the W. Indian Ocean; Point Quobba northwards; E. Indian Ocean and W. Pacific; to 43 cm; 2.154 kg. ★★★★

730 BLACKSPOT PIGFISH
Bodianus vulpinus (Richardson)
Inhabits mainly rocky reefs; male shown here distinguished by black blotch at base of middle dorsal spines and reddish margin on upper and lower edges of tail; female shown on Plate 47; Cape Naturaliste to Shark Bay; mainly W. Pacific; to 60 cm; 1.8 kg. ★★★★

938a BLUNT UNICORNFISH
Naso fageni Morrow
Inhabits coral reefs, often in schools; distinguished by lack of forehead spike (although large males may develop protruding snout) and pearl-white ring around tail base (rapidly disappears after death); Ningaloo Reef northwards; Indo-W. Pacific; to 80 cm. ★★

1003a PINKTAIL TRIGGERFISH
Melichthys vidua (Solander)
Inhabits coral reefs; similar to 1003 (p. 150), but distinguished by pale dorsal, anal, and tail fins; Ningaloo Reef northwards; Indo-C. Pacific; to 38 cm.

Triggerfishes are closely related to leatherjackets, a well known group in Australian seas. The Red-lined Triggerfish *(Balistapus undulatus)* inhabits coral reefs north of Ningaloo. It is one of many W.A. fishes that also occurs on the Great Barrier Reef. (See Plate 66, no. 1000 for additional information.)

The Green Moon Wrasse *(Thalassoma lutescens)* is common on coral and rocky reefs, ranging as far south as Rottnest Island. The male shown here displays a more brilliant colour pattern than the female. (See Plate 51, no. 780 for additional infomation.)

Hawfishes are small reef inhabitants which use their thickened pectoral rays to perch on coral lumps and rocks. The Ring-eyed Hawkfish *(Paracirrhites arcatus,* shown here, is occassionally sighted at Ningaloo Reef. (See Plate 45, no. 699 for additional information.)

The bright coloured Emperor Angelfish *(Pomacanthus imperator)* can produce a loud drumming noise that sometimes startles divers. This species is encountered on coral reefs north of Shark Bay. (See Plate 41, no. 645 for additional information.)

163

169

<table>
<tr><td></td><td>Species
No.</td><td></td><td>Species
No.</td></tr>
<tr><td>Flathead, Spiny</td><td>276</td><td>fuscoguttatus, Epinephelus</td><td>304</td></tr>
<tr><td>Flat-tail Mullet</td><td>709</td><td>fuscus, Bathygobius</td><td>878</td></tr>
<tr><td>Flat-tailed Longtom</td><td>172</td><td>holomelas, Atrosalarias</td><td>827</td></tr>
<tr><td>flavimarginatus, Gymnothorax</td><td>101</td><td>Pseudobalistes</td><td>1006</td></tr>
<tr><td>flavissimus, Forcipiger</td><td>630</td><td>Pseudochromis</td><td>334</td></tr>
<tr><td>flavomaculatus, Plectorhinchus</td><td>509</td><td>Fusigobius duospilus</td><td>887</td></tr>
<tr><td>flavomosaicus, Urolophus</td><td>68</td><td>Fusiler, Black-tipped</td><td>519</td></tr>
<tr><td>Flounder, Cockatoo</td><td>983</td><td>Blue</td><td>518</td></tr>
<tr><td>Deep-bodied</td><td>970</td><td>Fusiler, Red-bellied</td><td>517</td></tr>
<tr><td>Intermediate</td><td>972</td><td>fusovatus, Apogon</td><td>406</td></tr>
<tr><td>Large-toothed</td><td>968</td><td></td><td></td></tr>
<tr><td>Ocellated</td><td>984</td><td>G</td><td></td></tr>
<tr><td>Panther</td><td>975</td><td>gaimardi, Coris</td><td>747</td></tr>
<tr><td>Small toothed</td><td>971</td><td>Galeocerdo cuvier</td><td>33</td></tr>
<tr><td>Spiny-headed</td><td>973</td><td>Garfish, Barred</td><td>166</td></tr>
<tr><td>Three-spot</td><td>974</td><td>Buffon's</td><td>165</td></tr>
<tr><td>Twinspot</td><td>969</td><td>Long-finned</td><td>170</td></tr>
<tr><td>Flowery Cod</td><td>304</td><td>Quoy's</td><td>169</td></tr>
<tr><td>Flutemouth, Painted</td><td>204</td><td>Robust</td><td>167</td></tr>
<tr><td>Rough</td><td>203</td><td>Snub-nosed</td><td>164</td></tr>
<tr><td>Smooth</td><td>202</td><td>Three by Two</td><td>167</td></tr>
<tr><td>Flying Fish</td><td>163</td><td>Tropical</td><td>168</td></tr>
<tr><td>Foa brachygramma</td><td>371</td><td>gavialoides, Tylosurus</td><td>175</td></tr>
<tr><td>Forcipiger flavissimus</td><td>630</td><td>Gaza minuta</td><td>473</td></tr>
<tr><td>formosa, Cephalopholis</td><td>287</td><td>genivittatus, Lethrinus</td><td>531</td></tr>
<tr><td>formosus, Parioglossus</td><td>897</td><td>germaini, Omobranchus</td><td>835</td></tr>
<tr><td>fornasini, Lactoria</td><td>1027</td><td>Germain's Blenny</td><td>835</td></tr>
<tr><td>forsteri, Paracirrhites</td><td>698</td><td>geographicus, Anampses</td><td>724</td></tr>
<tr><td>Fourspot Wrasse</td><td>753</td><td>Gerres flamentosis</td><td>566</td></tr>
<tr><td>Fowleria aurita</td><td>373</td><td>oyena</td><td>567</td></tr>
<tr><td>variagatus</td><td>372</td><td>subfasciatus</td><td>568</td></tr>
<tr><td>fraenatus, Apogon</td><td>388</td><td>ghobban, Scarus</td><td>795</td></tr>
<tr><td>Freckled Anglerfish</td><td>153</td><td>Ghost Flathead</td><td>205</td></tr>
<tr><td>Hawkfish</td><td>698</td><td>Pipefish</td><td>211</td></tr>
<tr><td>Moray</td><td>108</td><td>Shark</td><td>72</td></tr>
<tr><td>Porcupinefish</td><td>1055</td><td>Giant Herring</td><td>73</td></tr>
<tr><td>frenatus, Scarus</td><td>793</td><td>Moray</td><td>103</td></tr>
<tr><td>Sufflamen</td><td>1009</td><td>Salmon Catfish</td><td>124</td></tr>
<tr><td>Frigate Mackerel</td><td>962</td><td>Seapike</td><td>721</td></tr>
<tr><td>Fringe-eyed Flathead</td><td>273</td><td>Threadfin</td><td>718</td></tr>
<tr><td>Fringe-finned Trevally</td><td>414, 445</td><td>Trevally</td><td>429</td></tr>
<tr><td>Fringe-lipped Snake-eel</td><td>113</td><td>gibbosa, Sardinella</td><td>89</td></tr>
<tr><td>Frogfish, Banded</td><td>146</td><td>gibbosus, Plectorhinchus</td><td>511</td></tr>
<tr><td>Dahl's</td><td>143</td><td>Tetrasomus</td><td>1033</td></tr>
<tr><td>Ocellated</td><td>147</td><td>gibsoni, Kyphosus</td><td>595</td></tr>
<tr><td>Three-spined</td><td>145</td><td>Giraffe Eel</td><td>100</td></tr>
<tr><td>Western</td><td>144</td><td>Girdled Goby</td><td>898</td></tr>
<tr><td>fronticinctus, Cantherhines</td><td>1014</td><td>Reef Eel</td><td>95</td></tr>
<tr><td>Frostback Cod</td><td>312</td><td>Gizzard Shad</td><td>83</td></tr>
<tr><td>fucata, Archamia</td><td>410</td><td>gladius, Xiphias</td><td>946</td></tr>
<tr><td>fulviflamma, Lutjanus</td><td>500</td><td>Glassy Bigeye</td><td>360</td></tr>
<tr><td>fulvoguttatus, Carangoides</td><td>420</td><td>Bombay Duck</td><td>142</td></tr>
<tr><td>fumea, Chromis</td><td>671</td><td>glauca, Prionace</td><td>36</td></tr>
<tr><td>furcosus, Nemipterus</td><td>542</td><td>glaucoparieus, Acanthurus</td><td>927</td></tr>
<tr><td>fuscescens, Siganus</td><td>921</td><td>Glaucosoma burgeri</td><td>347</td></tr>
</table>

	Species No.
murdjan, Myripristis	185
muricatum, Bolbometopon	788
Murion Pipefish	217
Muzzled Blenny	838
myops, Trachinocephalus	138
Myrichthys colubrinus	115
Myripristis adustus	181
hexagonatus	182
kuntee	183
melanostictus	184
murdjan	185
Myxus elongatus	712

N

	Species No.
nagasakiensis, Pentapodus	556
Pomacentrus	687
Naked-headed Catfish	126
Seabream	523
nalua, Ambassis	345
Narcine westaustraliensis	54
narinari, Aetobatus	71
Narrow Sawfish	44
Narrow-banded Batfish	605
Sergeant-Major	651
Narrow-barred Grubfish	809
Spanish Mackerel	954
Narrow-lined Cardinalfish	410
Toadfish	1037
Naso brevirostris	936
fageni	938a
tuberosus	937
unicornis	938
nasus, Rhynchostracion	1032
natans, Parapegasus	207
Naucrates ductor	444
naucrates, Echeneis	470
navarchus, Pomacanthus	644
Nebrius ferrugineus	3
nebulosa, Echidna	94
Parapercis	813
nebulosus, Halichoeres	760
Lethrinus	528
Yongeichthys	910
Zenopsis	198
Nebulous Wrasse	760
Needleskin Queenfish	450
Negaprion acutidens	35
negrosensis, Macrophargngodon	771
Nematalosa come	86
nematophorus, Nemipterus	548
Symphorus	489
nematophthalmus, Cymbacephalus	273
nematophorus, Nemipterus	548
nematopus, Nemipterus	552

	Species No.
Nemipterus baliensis	539
bathybius	540
celebicus	541
furcosus	542
hexodon	543
isacanthus	544
japonicus	545
mesoprion	546
nematophorus	548
nematopus	552
peronii	549
sp.	550
virgatus	551
zysron	547
Neoaploactis tridorsalis	265
neoguinaica, Albula	**74**
Neon Damsel	683
Threefin	860
Neoglyphidodon melas	**660**
nigroris	**661**
Neoniphon sammara	186
Neopomacentrus azysron	658
cyanomos	656
filamentosus	657
Netted Lizardfish	136
nichofii, Aetomyleus	70
nieuhofi, Triacanthus	994
niger, Melichthys	1003
Parastromateus	446
nigrescens, Halichoeres	758
nigricans, Stegastes	691
nigripinnis, Apogon	397
Polydactylus	717
nigrofasciata, Seriolina	460
nigrofuscus, Acanthurus	932
nigroocellatus, Istigobius	894
nigropunctatus, Arothron	1042
*nigroris, **Neoglyphidodon***	661
nigrosensis, Macropharyngodon	
niphonia, Pristigenys	359
Northern Bluefin Tuna	951
Dragonet	871
Pilchard	79
Sand Flathead	277
Scorpionfish	252
Slender Bullseye	591
Threadfin	715
Whiting	357
Wobbegong	15
North-West Black Bream	535
Blowfish	1050
Cod	308
Ruffe	990
Snapper	528
Notched Threadfin-Bream	549

<table>
<tr><td></td><td>Species
No.</td><td></td><td>Species
No.</td></tr>
</table>

Species
No.

196

Total of at least 175 species in 6 dives, 1 snorkel

Total of at least 175 species in 6 dives, 1 snorkel